I0504054

The PHYSICAL CONSTANTS

Back to Basics

By

Keith Dixon-Roche

Keith Dixon-Roche © 2017 to 2026

The PHYSICAL CONSTANTS

Back to Basics

Published by CalQlata
info@CalQlata.com

First published January 2019
Final publication April 2026

The PHYSICAL CONSTANTS

Contents

Preface

Since venturing into this mathematical field, I was struck by the apparent confusion in all physical constants and their units. Life would be so much easier if we could *cross-calculate* between the physical properties and scientific fields.

Properties such as permittivity, magnetic constant, electricity, etc. are far too obscure. Planck's universal energy constant (h) actually has incorrect units. The magnetic field constant (B) is the reciprocal of the relative charge capacity (RC) and should therefore be redundant. Heat capacity coefficients exist but there are no such coefficients for charge, etc.

It never occurred to me that a satisfactory resolution was possible until I began my work on Newton's laws of orbital motion as they apply to the atom. Just as it never occurred to me that everything in nature is simply electrical and magnetic charge and energy.

Throughout my life, every publication I referred to offers approximate (and frequently incorrect) values for all constants. Some of the more overt publications actually apply '±' tolerances in brackets to show how clever they are! In fact, *genuine* constants should be absolute and inflexible just as is the case for all of the mathematical laws of natural science.

I have always believed that all physical constants must apply equally to everything in nature, encompassing all scientific fields. This view has now turned out to be true. There is indeed one simple solution.

The historical units can all be resolved accurately from only six primary values (four constants and two ratios), all of which are accurately known. We no longer need to qualify constants with '±'. It is for this reason that I have derived values for all such constants accurate to 15 significant figures, except where absolute accuracy requires less. I leave it up to you to round them off if preferred. Remember, approximations frequently lead to inaccuracy and faulty conclusions.

However, if we go back to basics, and reset the four basic constants (magnetic & electrical charge, time and distance) to unity at the neutronic condition, there are now only three constants, all of which are ratios, and every physical constant we know today can be accurately derived from just these three.

Chapter 2 of this book has therefore been divided into two parts (A & B) to differentiate between the values (and their units) as we understand them today and those based upon the neutron:

Part A: of Chapter 2 applies to everyday calculations.

It provides all the constants and their formulas based upon historical values and units that we've been taught in schools and universities; Coulomb, kilogram, metre and second. These values and their units were established by physicists and administrators that were not aware of the true nature of particle physics and therefore do not correspond to reality.

Part B: of Chapter 2 is exactly the same (in all respects) as Part A except for the formulas and values, which are here based upon the neutron. Its units are therefore; Coulomb, Gilbert, metre and second; the values of which are reset with the neutronic condition representing unity. This part of Chapter 2 has been provided for information only, but it constitutes the future; *actual reality*.

Apart from my discovery of the nature of the neutron, this compilation of physical constants is the work of which I am most delighted, because, like the neutron, it is all mine and it will make my future work so much easier.

Keith Dixon-Roche 2026

1   Introduction

The physical constants are the most important tool in the evaluation and description of natural laws. And the most important natural law is; 'everything in the universe is *energy*', which is currently described thus:

Dynamic (kinetic): $KE = \frac{1}{2}.m.v^2$

static (potential): $PE = m.a.R$ $(a = v^2/R)$

Where 'm' above represents what is currently referred to as mass, but is actually magnetic charge (chapter 3.2). However, I realise that this is a giant leap of faith for most of us to take along with all the other revelations described here, so I shall continue in this book to refer to 'm' as mass; but to me; 'm' will always represent magnetic charge.

Temperature is not an *actual* unit of measurement; it is a contrived property to define the energy of the proton-electron pairs radiating the EME we sense as heat. This can be demonstrated by the replacement of the age-old method for calculating gas pressure; "$p.V = n.R_i.\underline{T}$" with an alternative calculation method using the energy of a proton-electron (chapter 3.5.3).

In fact, all physical properties can be established using the kinetic energy of electrons, you simply need to know how.

My values and units are based upon four primary constants (mass [magnetic charge], electrical charge, distance and time), two primary ratios (static and dynamic) and a particle constant (chapter 3.6.1).

None of this work would have been possible without the discovery of the neutronic radius (R_n), which is the primary unit of distance. When united with the neutronic period (t_n), mass (m) and electrical charge (e), these constants and their formulas provide us with all we need to accurately calculate all physical constants and variables.

All properties can be established simply by multiplying and/or dividing various appropriate constants to achieve the value and units you need.

The level of accuracy established for the primary constants has permitted the mathematical prediction of all physical properties with absolute accuracy and reliability; there is no need for '±' tolerances.

2   The Physical Constants

References within Chapter 2 apply only to the Part concerned (A or B).
For example;
A reference to Table '7' in Part A applies only to Table A7 in Part A,
and;
a reference to Table '7' in Part B applies only to Table B7 in Part B.

All new constants (unknown until now) are highlighted in **bold text**.

The PHYSICAL CONSTANTS

2.1 Symbols

The following is a list of the symbols that can be found in both Parts A & B of this chapter. The Table(s) in the list refer to the last character(s) in the title, e.g.; 4 = Table 2.4

Symbol	Description	Table(s)
a_o	Rydberg radius	2a
A	EME amplitude	8.1, 8.2
I	electrical current	7
c	speed of electro-magnetic radiation	2b
$c_?$	specific heat capacity	4
$C_?$	heat capacity	4
e	elementary charge unit	1
e_n	proton neutronic charge	2a
e	natural logarithm	2b
E	energy	
F	Farad	7
F	force	
G	Newton's gravitational constant	2a
h	Planck's constant	2a
h'	Planck's constant (modified)	2a
H	Henry	7
k	Coulomb's constant	2a
k'	Coulomb's constant (modified)	2a
k_B	Boltzmann's constant	3
K	constant of proportionality (proton-electron pair)	2b
$m_?$	mass	8.1, 8.2
m_u	unit mass of ultimate density	2b
$N_?$	microstate	6
N_A	Avogadro's number	3
$q_?$	specific charge capacity	5
$Q_?$	charge capacity	3, 5
$r_?$	particle radii	8.1, 8.2
$R_?$	orbital radii	2b
$R_?$	gas constant	3

Table 2.1a

The PHYSICAL CONSTANTS

Symbol	Description	Table(s)
R_a	specific gas constant	4
R_i	ideal gas constant	3
RAC	relative atomic charge	5
RAM	relative atomic mass	4
RC	relative charge capacity	3
R_n	neutronic radius	1
R_T	gas constant (temperature dependent)	3
R_∞	Rydberg's wave number	2a
R_γ	Rydberg's universal constant (energy)	2a
R_r	EME wavelength constant (orbital radius)	2b
t	orbital periods	8.1, 8.2
t_n	neutronic orbital period	1
$\underline{T}_?$	temperatures	2b
$\underline{T}_n$	neutronic temperature	2b
$v_?$	orbital velocities	2b
V	electrical voltage	7
V	particle volumes	8.1, 8.2
X	heat coefficient (velocity)	2a
X_R	heat coefficient (orbital radius)	2a
Y	temperature coefficient	2a
ϵ_e, ϵ_0	particles per unit mass of ultimate density	2b
$\varepsilon, \varepsilon_p$	permittivity of a vacuum	2a
λ	wavelength	8
μ, μ_0	magnetic constant	2a
ρ_u	ultimate density	2b
Σ	particle constant	1
φ	coupling ratio	2a
γ	Rydberg's constant	2a
κ	EME wavelength (constant)	2b
ζ_m	static ratio	1
ζ_v	dynamic ratio	1

Table 2.1b

6

The PHYSICAL CONSTANTS

Symbol	Description
Suffix:	
e	electron
p	proton
p	constant pressure (heat & charge capacity)
t	constant temperature (heat & charge capacity)
u	ultimate
v	constant volume (heat & charge capacity)
Atomic: temperature, velocity & orbital radius:	
c	cold
m	mean Planck value
n	neutronic
o	minimum Planck value
Modifier:	
N	Newton
P	Planck
Table 2.1c	

The PHYSICAL CONSTANTS

Chapter 2

Part A

The constants we understand today

The PHYSICAL CONSTANTS

2A.1   Primary Constants

All universal properties can be defined from only *four* primary constants, *two* ratios and a particle constant (Σ), all of which are listed below:

Symbol	Value	Units
m_e	**9.1093897E-31**	kg
electron mass (magnetic charge)		
e	**1.60217648753E-19**	C
electron electrical charge	(chapter 3.6.7)	
R_n	**2.81793795383896E-15**	m
neutronic radius	(chapter 3.6.10)	
t_n	**5.90596121302193E-23**	s
neutronic period	(chapter 3.6.10)	
ξ_m	**1836.15115053207**	
static ratio	(chapter 3.6.15)	
ξ_v	**1722.0458764934**	
dynamic ratio	(chapter 3.6.15)	
Σ	**3E-91** (exact)	m^6
particle constant	(chapter 3.6.1)	
Table A1		

Whilst *mass* is actually *magnetic charge* and *gravity* is actually *magnetism*, I shall refer to *mass* and *gravity* in this book in order to minimise confusion.

2A.2   Principal Constants

Symbol	Formula	Value	Units
G	$a_o.c^2 / m_u$	6.67359232004334E-11	$m^3 / s^2.kg$
Newton's gravitational constant			chapter 3.6.3
k	$1/\varepsilon_o$	8.98755184732667E+09	$J.m / C^2$
Coulomb's constant			chapter 3.6.5
k'	$k.RC^2$	2.78024810626745E+32	$m^3 / kg.s^2$
Coulomb's constant (*modified*)			chapter 3.6.5
φ	$G.m_e.m_p / k.e^2$	4.40742111792334E-40	
coupling ratio			
μ_o	$R_n.m_e/e^2$	1E-07	$kg.m / C^2$
Henry's magnetic constant			
μ	$4\pi.\mu$	1.25663706143592E-06	$kg.m / C^2$
magnetic constant (*spherical*)			
ε_o	$1 / \mu_o.c^2$	1.11265004863082E-10	$C^2 / J.m$
permittivity constant			
ε	$1 / \mu.c^2$	8.85418775855161E-12	$C^2 / J.m$
permittivity constant			
h	$\frac{1}{2}.m_e.c.\xi_v . R_n$	6.62607174469163E-34	$kg.m^2/s$
Planck's constant			
h'	$\frac{1}{2}.m_e.c^2 . R_n$	1.15353857232684E-28	$J.m$
Planck's constant (*modified*)			
$\hbar$	$h/2\pi$	1.054572071449210E-34	$kg.m^2/s$
Planck's constant (Dirac)			
γ	$(\xi_v / 4\pi)^2$	18778.8808461551	
Rydberg's constant			
a_o	$R_n.\gamma$	5.2917721067E-11	m
Rydberg's radius			
R_∞	$1 / a_o.\xi_v$	1.09737269561359E+07	/m
Rydberg's wave number			
R_γ	$\frac{1}{2}.m_e.c^2 / \gamma$	2.17987197684936E-18	J
Rydberg's electron energy constant			
X	$\underline{T}_n/c^2$	6.9353271647894E-09	$K.s^2/m^2$
heat transfer coefficient (*velocity*)			
X_R	$\underline{T}_n.R_n$	1.75646616508035E-06	K.m
heat transfer coefficient (*radial*)			
Y	$\sqrt[3]{[\frac{1}{2}.\xi_v]}$	9.51345439232503	
temperature coefficient			
e_n	$m_p.RC$	2.94183820093364E-16	C
proton charge (*neutronic*)			
Table A2a			

13

The PHYSICAL CONSTANTS

Atomic property constants (chapter 8.1; particle properties):

Symbol	Formula	Value	Units
ρ_u	$m_e.\sqrt{[\xi_m/\Sigma]}$	7.1266079635045E+16	kg/m³
ultimate density			
m_u	$\rho_u/1$	7.1266079635045E+16	kg
unit mass of ultimate density			
R_x	X_R/T_x	8.59854098572228E-07	m
Cold orbital radius			
R_o	$R_n.\xi_v{}^2$	8.3564315638157E-09	m
Planck minimum orbital radius			
R_m	$R_n.\xi_v$	4.85261843362263E-12	m
Planck mean orbital radius			
R_r	R_o/κ	8.40016460895157E-11	m
EME orbital radius constant			
v_x	$\sqrt{[T_x/X]}$	17162.2425219270	m/s
electron cold velocity			
v_o	c/ξ_v	174090.866621084	m/s
electron minimum orbital velocity (Planck)			
v_m	$c/\sqrt{\xi_v}$	7224342.80705004	m/s
electron mean orbital velocity (Planck)			
c	$2\pi.R_n/t_n$	299792459	m/s
electron neutronic velocity			
T_x	$X.(c / Y.\xi_m)^2$	2.04274907568265	K
cold temperature			
T_o	$X.v_o{}^2$	210.193328535837	K
Planck minimum temperature		chapter 3.6.6; Table 3.6.6-1	
T_m	$X.v_m{}^2$	361962.554671561	K
Planck mean temperature		chapter 3.6.6; Table 3.6.6-1	
T_n	$X.c^2$	623316124.717179	K
neutronic temperature		chapter 3.6.6; Table 3.6.6-1	
h_e	$c.R_n$	8.4479654849081E-07	m²/s
Newton's constant of motion (*electron*)			
κ	$(2\pi)^2 . 2^{4/3}$	99.4793787125405	
EME wavelength (constant)			
K	$t_n{}^2/R_n{}^3$	0.15587874533403	s²/m³
constant of proportionality (proton-electron pair)		chapter 3.6.14	
e	exp(1)	2.71828182845905	
Natural logarithm			
ϵ_e	m_e/ρ_u	7.82336489952175E+46	
ϵ_p	m_p/ρ_u	4.26074122343073E+43	
number of particles in a unit mass of ultimate density			
Table A2b			

2A.3   Universal Heat & Charge Capacities

Symbol	Formula	Value	Units
RC	e/m_e	1.75881869180545E+11	C/kg
Relative charge capacity			
k_B	$m_e / X.Y$	1.38065156E-23	J/K
Boltzmann's constant			
N_A	$2 / (m_e + m_p + m_n).1000$	5.97538412973187E+23	/mole
Avogadro's number (based upon C_{12})			
R_i	$k_B.N_A$	8.24992342031355	J / K.mol
Ideal gas constant; Avogadro's number (*based upon C_{12}*)			
R_e	$c_{pe}.RAM_e$	20.6248085507841	J / K.mol
Gas constant; R_i (electron)			
R_p	$c_p.RAM$	20.6248085507839	J / K.mol
Gas constant; R_i multiplied by 2.5 (*based upon C_{12}*)			
R_T	$RAC.q_p.Ln(\underline{T})$ $R_i.Ln(N_t)$		J / K.mol
Gas constant ($R_T = R_i$ when $N_t = e$ & $\underline{T} = 1.49182469764127$ K)			
C_t	$m.c_t$		J/K
Heat capacity (*constant temperature*)			
C_V	$m.c_V$		J/K
Heat capacity (*constant volume*); C_t multiplied by 1.5			
C_p	$m.c_p$		J/K
Heat capacity (*constant pressure*); C_t multiplied by 2.5			
Q_t	$e.q_t$		J/K
Charge capacity (*constant temperature*); also equal to **R** & C_t			
Q_V	$e.q_V$		J/K
Charge capacity (*constant volume*); Q_t multiplied by 1.5; also equal to C_V			
Q_p	$e.q_p$		J/K
Charge capacity (*constant pressure*); Q_t multiplied by 2.5; also equal to C_p			
Table A3			

2A.4 Specific Heat Capacities (particles)

Symbol	Formula	Value	Units
RAM_e	$m_e . N_A$	5.4432102644923E-07	kg/mol
Relative atomic mass of an electron			
RAM_p	$m_p . N_A$	9.99455678973551E-04	kg/mol
Relative atomic mass of a proton (also the RAM of an hydrogen atom)			
R_{ae}	$R_i / RAM_e = k_B / m_e$	1.51563563034308E+07	J / kg.K
Specific gas constant for an electron			
R_{ap}	$R_i / RAM_p = k_B / m_p$	8.25441647276088E+03	J / kg.K
Specific gas constant for a proton			
c_{et}	k_B / m_e	1.51563563034305E+07	J / kg.K
Specific heat capacity for the electron (*constant temperature*)			
c_{eV}	$1.5 . c_{et}$	2.27345344551458E+07	J / kg.K
Specific heat capacity for the electron (*constant volume*)			
c_{ep}	$c_{et} + c_{eV}$	3.78908907585763E+07	J / kg.K
Specific heat capacity for the electron (*constant pressure*)			
c_{pt}	k_B / m_p	8.25441647276074E+03	J / kg.K
Specific heat capacity for the proton (*constant temperature*)			
c_{pV}	$1.5 . c_{pt}$	1.23816247091411E+04	J / kg.K
Specific heat capacity for the proton (*constant volume*)			
c_{pp}	$c_{pt} + c_{pV}$	2.06360411819018E+04	J / kg.K
Specific heat capacity for the proton (*constant pressure*)			
C_t	$m_e . c_{et}$ $m_p . c_{pt}$	1.38065156E-23	J/K
Heat capacity (*constant temperature*); equal to **R & Q_t**			
C_V	$m_e . c_{eV}$ $m_p . c_{pV}$	2.07097734E-23	J/K
Heat capacity (*constant volume*); equal to **Q_V**			
C_p	$m_e . c_{ep}$ $m_p . c_{pp}$	3.4516289E-23	J/K
Heat capacity (*constant pressure*); equal to **Q_p**			

Table A4

The PHYSICAL CONSTANTS

2A.5   Specific Charge Capacities (particles)

Symbol	Formula	Value	Units
RAC_e	$e.N_A$	96485.3317942158	C/mol
Relative atomic charge of an electron (also equal to the Farad)			
RAC_p	$e'.N_A$	1.77161652983418E+08	C/mol
Relative atomic charge of a proton (also the RAC of an hydrogen atom)			
R_{ce}	R_i / RAC_e	8.61735002820125E-05	J / C.K
Specific gas constant for an electron			
R_{cp}	R_i / RAC_p	4.69315939796359E-08	J / C.K
Specific gas constant for a proton			
q_{et}	k_B / m_e	8.61735002820123E-05	J / C.K
Specific charge capacity for the electron (*constant temperature*)			
q_{eV}	$1.5 . q_{et}$	1.29260250423019E-04	J / C.K
Specific charge capacity for the electron (*constant volume*)			
q_{ep}	$q_{et} + q_{eV}$	2.1543375070503E-04	J / C.K
Specific charge capacity for the electron (*constant pressure*)			
q_{pt}	k_B / e'	4.69315939796358E-08	J / C.K
Specific charge capacity for the proton (*constant temperature*)			
q_{pV}	$1.5 . q_{pt}$	7.0397390969454E-08	J / C.K
Specific charge capacity for the proton (*constant volume*)			
q_{pp}	$q_{pt} + q_{pV}$	1.1732898494909E-07	J / C.K
Specific charge capacity for the proton (*constant pressure*)			
Q_t	$e.q_{et}$ $e'.q_{pt}$	1.38065156E-23	J/K
Charge capacity (*constant temperature*); equal to **R** & C_t			
Q_V	$e.q_{eV}$ $e'.q_{pV}$	2.07097734E-23	J/K
Charge capacity (*constant volume*); equal to C_V			
Q_p	$e.q_{ep}$ $e'.q_{pp}$	3.4516289E-23	J/K
Charge capacity (*constant pressure*); equal to C_p			
Table A5			

2A.6   Microstates

Symbol	Formula	Value	Units
N_t	$\exp(c_p \cdot L_n(\underline{T}) / R_a)$ $\exp(q_p \cdot L_n(\underline{T}) / R_a)$ $\exp(2.5 \cdot Ln(\underline{T}))$		
Microstate (constant *temperature*)			
N_V	c_v / R_a q_v / R_a		
Microstate (constant *volume*); N_t multiplied by 1.5			
N_p	c_p / R_a q_p / R_a		
Microstate (constant *pressure*), N_t multiplied by 2.5			
Table A6			

The PHYSICAL CONSTANTS

2A.7   Electricity

Apart from the Farad, no values are provided for the following electrical properties because the Amp, Volt, Ohm and Henry are now redundant.

Symbol	Formula	Value	Units
I	$e.f$		C/s
Electrical current (Coulomb flow-rate)			
V	PE/e		J/C
Electrical voltage (potential energy per coulomb)			
Ω	$V/I = PE / f.e^2$		$J.s/C^2$
Electrical resistance (momentum over distance per Coulomb squared)			
H	$\mu.R$		$kg.m^2/C^2$
Henry; unit of mutual inductance (R = distance of induction)			
F	$e.N_A$	96485.3317942156	C/mol
Farad; unit of electrostatic capacitance (equal to RAC_e)			
P	$V.I = PE.f$		J/s
Power (Watt)			
Table A7			

PE is the potential energy between a proton and its orbiting electron

2A.8   Newton vs Planck

Newton's atom is the real one; the one we see, hear and feel all around us. Planck's atom is theoretical, it does not exist. It is based upon his three constants; time (t^P), length (A^P) and mass (m^P). See Table A8 below.

It should be noted here that Planck's length is actually an orbital radius (R) and not a wavelength (λ) as depicted below. The reason being; force (F) is Energy divided by length, and a length in such a formula can only be an orbital radius.

The atomic particles in Planck's atom are identical in size; his electron is exactly the same, in all respects, as his proton. The only consistency between the two atoms (Newton's and Planck's) is that the product of the particle volumes in each atom is identical; $V_e.V_p = V^P.V^P = \Sigma$

It is the comparison between these two atoms that has given us the ability to solve Newton's gravitational constant (G; see Chapter 3.6.3)

The following Tables (A8a to A8c) provide comparative results for three atomic variations in which the particle properties are defined thus:

Newton atom; comprising Newtonian particles (Table A8a)
$t = a_o/c$; $A = a_o$; $m = \rho_u$; $E = m_u.c^2$; $F = E/a_o$

Planck^N atom; comprising Newtonian particles (Table A8a)
t^P; A^P; m^P; E^P; F^P (Table A8b)

Planck^P atom; comprising Planck particles (Table A8c), and calculated using a revised version of his own constant; ħ:
$ħ = \sqrt{[\, \pi.m.e^2.a_o / \varepsilon_o \,]} = 7.99473592559180E\text{-}16$ kg.m^2/s
in which:
$a_o = A^P / 4\pi^2$
$e = \sqrt{[\, G.m^{P2} / k.\varphi \,]}$
$m = m^P$

Note: 'c', 'ε$_o$', 'G', 'φ' and 'k' are equal in both Newton and Planck atoms

wton atom compared with a Planck^N atom:

	Newton Atom (A)	Planck^N Atom (B)	A/B
t	1.76514516887831E-19	5.39096122598359E-44	3.27426797353056E+24
A	5.2917721067E-11	1.61616952231128E-35	3.27426797353056E+24
m	7.1266079635045E+16	2.1765500017459E-08	3.27426797353056E+24
E	6.40507585675678E+33	1.95618559889902E+09	3.27426797353056E+24
F	1.21038391820525E+44	1.21038391820525E+44	1
Table A8a			

Newton atom compared with a Planck^P atom:

	Newton Atom (A)	Planck^P Atom (B)	A/B
t	1.76514516887831E-19	1.48432887846076E-34	1.18918737922067E+15
A	5.2917721067E-11	4.44990604438464E-26	1.18918737922067E+15
m	7.1266079635045E+16	59.9283854507006	1.18918737922067E+15
E	6.40507585675678E+33	5.38609471364748E+18	1.18918737922067E+15
F	1.21038391820525E+44	1.21038391820525E+44	1
Table A8b			

Planck^N atom compared with a Planck^P atom:

	Planck^N Atom (A)	Planck^P Atom (B)	A/B
t	5.39096122598359E-44	1.48432887846076E-34	2.75336589568949E+09
A	1.61616952231128E-35	4.44990604438464E-26	2.75336589568949E+09
m	2.1765500017459E-08	59.9283854507006	2.75336589568949E+09
E	1.95618559889902E+09	5.38609471364748E+18	2.75336589568949E+09
F	1.21038391820525E+44	1.21038391820525E+44	1
Table A8c			

Note: 3.27426797353056E+24 ÷ 1.18918737922067E+15 = 2.75336589568949E+09

26

2A.8.1   Newton's Atom (particle properties)

The following Table provides the properties of particles in the atom that exists in nature.

Symbol	Formula	Value	Units
m_e			
electron mass (Table A1)			
m_p	$\xi_m.m_e$	1.67262163783E-27	
proton mass			
m_n	$m_e + m_p$	1.6735325768E-27	kg
neutron mass			
V_e	m_e / ρ_u	1.27822236702922E-47	m^3
electron volume			
V_p	m_p / ρ_u	2.34700946985653E-44	m^3
proton volume			
V_n	m_n / ρ_u	2.34828769222356E-44	m^3
neutron volume			
r_e	$\sqrt{[\,3.V_e / 4\pi\,]}$	1.45046059426276E-16	m
electron radius			
r_p	$\sqrt{[\,3.V_p / 4\pi\,]}$	1.77613270336827E-15	m
proton radius			
r_n	$\sqrt{[\,3.V_n / 4\pi\,]}$	1.77645508248591E-15	m
neutron radius			
t^N	a_o / c	1.765145168878310E-19	s
Newton time			
A^N	a_o	5.291772106700000E-11	m
Newton EME amplitude			
m^N	$a_o.c^2 / G$	7.126607963504500E+16	kg
Newton mass			
E^P	$m^N.c^2$	6.405075856756780E+33	J
Newton energy			
F^N	E^P / λ^N	1.210383918205250E+44	N
Newton force			
ρ_u	$m_e . \sqrt{[\xi_m/\Sigma]}$	7.1266079635045E+16	kg/m^3
The ultimate density of Newton's atomic particles			
Table A8.1			

2A.8.2 Planck's Atom (particle properties)

The following Table provides the properties of particles in a fictitious atom that has been constructed using Planck's values; t^P, A^P, m^P

Symbol	Formula	Value	Units
m_e^P	m^P	2.1765500017459E-08	kg
electron mass			
m_p^P	m^P	2.1765500017459E-08	kg
proton mass			
m_n^P	N/A	N/A	N/A
neutron mass			
V_e^P	$\sqrt{\Sigma}$	5.47722557505166E-46	m^3
electron volume			
V_p^P	$\sqrt{\Sigma}$	5.47722557505166E-46	m^3
proton volume			
V_n^P	N/A	N/A	N/A
neutron volume			
r_e^P	$\sqrt{[\,3.V_e^P/4\pi\,]}$	5.07563837996471E-16	m
electron radius			
r_p^P	$\sqrt{[\,3.V_p^P/4\pi\,]}$	5.07563837996471E-16	m
proton radius			
r_n^P	N/A	N/A	N/A
neutron radius			
t^P	$\sqrt{[\,\hbar.G/c^5\,]}$	5.39096122598359E-44	s
Planck time			
A^P	$\sqrt{[\,\hbar.G/c^3\,]}$	1.61616952231128E-35	m
Planck EME amplitude			
m^P	$\sqrt{[\,\hbar.c/G\,]}$	2.1765500017459E-08	kg
Planck mass			
E^P	$m^P.c^2$	1.95618559889902E+09	J
Planck energy			
F^P	c^4/G	1.21038391820525E+44	N
Planck force			
ρ_u^P	m^P/V_e^P	3.97381844498046E+37	kg/m^3
The ultimate density of Planck's atomic particles			
Table A8.2			

Chapter 2
Part B

The constants of the future

The PHYSICAL CONSTANTS

The revelation from unifying the physical constants at the neutronic condition is that every physical constant (without exception) can be calculated using just three ratios:

Circular ratio: $\pi = 3.14159265358979$

Static ratio: $\xi_m = 1836.15115053207$

Dynamic ratio: $\xi_v = 1722.0458764934$

Note: … and the magnetic constant changes from:
$\mu = 1\text{E-}07 \text{ kg.m/C}^2$ as it is today; to $\mu = 1 \text{ G.m/C}^2$ (unity!)

The constants in all the other Tables (2B.4 to 2B.7) can be calculated using the values established from 2B.3

Using these neurotically unified values and units;
Coulomb's constant becomes:
$k = \mu.c^2 = 1 \times (2\pi)^2 = 39.4784176043574 \text{ \{G.m}^3 / C^2.s^2\}$
Note; c = 2π m/s

'k' comprises all the units: electrical & magnetic charge, distance and time; i.e. Coulomb's becomes the defining constant. It contains everything.

Whilst Newton defined atomic orbits, Coulomb defined the atom's electrical and magnetic properties; what makes them work.

2B.1   Primary Constants

All universal properties can be defined from only *four* primary constants, *two* ratios and a particle constant (Σ), all of which are listed below:

Symbol	Value	Units
m_e	1	G
electron mass (magnetic charge)		
e	1	C
electron electrical charge		
R_n	1	m
neutronic radius		
t_n	1	s
neutronic period		
ζ_m	1836.15115053207	
static ratio		
ζ_v	1722.0458764934	
dynamic ratio		
Σ	3E-91	m^6
particle constant		
Table B1		

Whilst *mass* is actually *magnetic charge* and *gravity* is actually *magnetism*, I shall refer to *mass* and *gravity* in this book in order to minimise confusion.

The value of the magnetic charge (m) is set to the electrical charge of an electron and the unit Gilbert (G) has been selected for its unit of measurement.

2B.2 Principal Constants

Symbol	Formula	Value	Units
G	$\varphi \cdot k/\xi_m$	9.47623573370968E-42	$m^3 / s^2.G$
Newton's gravitational constant			
k	$(2\pi)^2$	39.4784176043574	$J.m / C^2$
Coulomb's constant [for an electron]			
k'	$k.RC^2$	39.4784176043574	$m^3 / s^2.kg$
Coulomb's constant (*modified*)			
φ	$(\xi_v / 4\pi)^2 \cdot \sqrt{[\sum.\xi_m]}$	4.40742111792334E-40	
Coupling Ratio			
μ_o	$R_n.m_e/e^2$	1	$G.m / C^2$
magnetic constant (fundamental)			
μ	4π	12.5663706143592	$G.m / C^2$
magnetic constant (spherical)			
ε_o	$1 / (2\pi)^2$	2.53302959105844E-02	$C^2 / J.m$
Permittivity of a vacuum (e.g. within an atom)			
ε	$1 / 2.(2\pi)^3$	2.01572090207497E-03	$C^2 / J.m$
Permittivity of a vacuum (e.g. within an atom)			
h	$\pi.\xi_v$	5409.96667473626	$G.m^2/s$
Planck's constant (resolved into its component parts)			
h'	$2.\pi^2$	19.7392088021787	$J.m$
Modified Planck's constant			
$\hbar$	$h / 2\pi$	861.0229382467	$G.m^2/s$
Plank's constant (Dirac)			
γ	$(\xi_v / 4\pi)^2$	18778.8808461551	
Rydberg's constant			
a_o	γ	18778.8808461551	m
Rydberg's radius			
R_∞	$(4\pi)^2 / \xi_v^3$	3.09232816847610E-08	/m
Rydberg's wave number			
R_γ	$2.(4.\pi^2/\xi_v)^2$	1.05113872141216E-03	J
Rydberg's universal constant for the energy of an electron			
X	$T_n / (2\pi)^2$	1.57887818849249E+07	$K.s^2/m^2$
heat transfer coefficient (velocity)			
X_R	T_n	6.23316124717178E+08	K.m
heat transfer coefficient (radial)			
Y	$\sqrt[3]{[½.\xi_v]}$	9.51345439232503	
Temperature coefficient			
e_n	ξ_m	1836.15115053207	C
Proton charge (neutronic)			
Table B2a			

Symbol	Formula	Value	Units
ρ_u	$\sqrt{[\xi_m/\Sigma]}$	7.82336489952175E+46	G/m^3
ultimate density			
m_u	$\sqrt{[\xi_m/\Sigma]}$	7.82336489952175E+46	G
unit mass of ultimate density			
R_x	T_x/X_R	8.59854098572228E-07	m
Cold orbital radius			
R_o	ξ_v^2	2.96544200074792E+06	m
Planck minimum orbital radius			
R_m	ξ_v	1722.0458764934	m
Planck mean orbital radius			
R_r	$\xi_v^2/$	29809.6152099721	m
EME orbital radius constant			
v_x	$\sqrt{[T_x/X]}$	5.18824113201723E-04	m/s
electron cold velocity			
v_o	$2\pi/\xi_v$	3.64867474957986E-03	m/s
electron minimum Planck orbital velocity			
v_m	$2\pi/\sqrt{\xi_v}$	0.15141102858523	m/s
electron mean Planck orbital velocity			
c	2π	6.28318530717959	m/s
electron neutronic velocity			
T_x	$X.(c/Y.\xi_m)^2$	2.04274907568265	K
cold temperature			
T_o	T_n/ξ_v^2	210.193328535837	K
Planck minimum temperature			
T_m	T_n/ξ_v	361962.554671561	K
Planck mean temperature			
T_n	$(2\pi)^2/Y.k_B$	623316124.717179	K
neutronic temperature			
h_e	2π	6.28318530717959	m^2/s
Newton's constant of motion (*electron*)			
κ	$24.(2\pi)^2/3$	99.4793787125405	
EME wavelength (constant)			
K		1	s^2/m^3
Constant of proportionality (proton-electron pair)			
e	exp(1)	2.71828182845905	
Natural logarithm			
ϵ_e	m_e/ρ_u	7.82336489952175E+46	
ϵ_p	m_p/ρ_u	4.26074122343073E+43	
number of particles in a unit mass of ultimate density			
Table B2b			

2B.3   Universal Heat & Charge Capacities

Symbol	Formula	Value	Units
RC	e/m_e	1	C/G
Relative charge capacity			
k_B	$2\pi / \sqrt[3]{[\tfrac{1}{2}.\xi_v]} / T_n$	1.35996633221504E-08	J/K
Boltzmann's constant			
NA	$1/(1 + \xi_m)/1000$	5.44321026449230E-07	/mol
Avogadro's number			
R_i	$k_B.N_A$	3.62383358142172E-15	J / K.mol
Ideal gas constant; Avogadro's number (original)			
R_e	$c_{pe}.RAM_e$	9.0595839535543E-15	J / K.mol
Gas constant; R_i (electron)			
R_p	$c_{pp}.RAM_p$	9.0595839535543E-15	J / K.mol
Gas constant; R_i (proton)			
R_T	$RAC.q_p.Ln(T)$ $R_i.Ln(N_t)$		J / K.mol
Gas constant ($R_T = R_i$ when $N_t = e$ & $T = 1.49182469764127$ K)			
C_t	$m.c_t$		J/K
Heat capacity (constant *temperature*)			
C_V	$m.c_V$		J/K
Heat capacity (constant *volume*); C_t multiplied by 1.5			
C_p	$m.c_p$		J/K
Heat capacity (constant *pressure*); C_t multiplied by 2.5			
Q_t	$e.q_t$		J/K
Charge capacity (constant *temperature*); also equal to **R** & C_t			
Q_V	$e.q_V$		J/K
Charge capacity (constant *volume*); Q_t multiplied by 1.5; also equal to C_V			
Q_p	$e.q_p$		J/K
Charge capacity (constant *pressure*); Q_t multiplied by 2.5; also equal to C_p			

Table B3

2B.4 Specific Heat Capacities (particles)

Symbol	Formula	Value	Units
RAM$_e$	$m_e.N_A$	5.44321026449230E-07	G/mol
Relative atomic mass of an electron			
RAM$_p$	$m_p.N_A$	9.99455678973551E-04	G/mol
Relative atomic mass of a proton (also the RAM of an hydrogen atom)			
R_{ae}	$R_i / RAM_e = k_B / m_e$	6.65753003344558E-09	J / G.K
Specific gas constant for an electron			
R_{ap}	$R_i / RAM_p = k_B / m_p$	3.62580718451005E-12	J / G.K
Specific gas constant for a proton			
c_{et}	k_B / m_e	6.65753003344558E-09	J / G.K
Specific heat capacity for the electron (constant *temperature*)			
c_{eV}	$1.5 . c_{et}$	9.98629505016837E-09	J / G.K
Specific heat capacity for the electron (constant *volume*)			
c_{ep}	$c_{et} + c_{eV}$	1.66438250836139E-08	J / G.K
Specific heat capacity for the electron (constant *pressure*)			
c_{pt}	k_B / m_p	3.62580718451005E-12	J / G.K
Specific heat capacity for the proton (constant *temperature*)			
c_{pV}	$1.5 . c_{pt}$	5.43871077676508E-12	J / G.K
Specific heat capacity for the proton (constant *volume*)			
c_{pp}	$c_{pt} + c_{pV}$	9.06451796127513E-12	J / G.K
Specific heat capacity for the proton (constant *pressure*)			
C_t	$m_e.c_{et}$ $m_p.c_{pt}$	6.65753003344566E-09	J/K
Heat capacity (constant *temperature*); equal to **R** & **Q$_t$**			
C_V	$m_e.c_{eV}$ $m_p.c_{pV}$	9.9862950501685E-09	J/K
Heat capacity (constant *volume*); equal to **Q$_V$**			
C_p	$m_e.c_{ep}$ $m_p.c_{pp}$	1.664382508361420E-08	J/K
Heat capacity (constant *pressure*); equal to **Q$_p$**			
Table B4			

2B.5 Specific Charge Capacities (particles)

Symbol	Formula	Value	Units
RAC_e	$e.N_A$	4.53600855374358E-05	C/mol
Relative atomic charge of an electron (also equal to the Farad)			
RAC_p	$e'.N_A$	0.0832879732477958	C/mol
Relative atomic charge of a proton (also the RAC of an hydrogen atom)			
R_{ce}	R_i / RAC_e	6.65753003344558E-09	J / C.K
Specific gas constant for an electron			
R_{cp}	R_i / RAC_p	3.62580718451005E-12	J / C.K
Specific gas constant for a proton			
q_{et}	k_B / m_e	6.65753003344558E-09	J / C.K
Specific charge capacity for the electron (constant *temperature*)			
q_{eV}	$1.5 . q_{et}$	9.98629505016837E-09	J / C.K
Specific charge capacity for the electron (constant *volume*)			
q_{ep}	$q_{et} + q_{eV}$	1.664382508361390E-08	J / C.K
Specific charge capacity for the electron (constant *pressure*)			
q_{pt}	k_B / e'	3.62580718451005E-12	J / C.K
Specific charge capacity for the proton (constant *temperature*)			
q_{pV}	$1.5 . q_{pt}$	5.43871077676508E-12	J / C.K
Specific charge capacity for the proton (constant *volume*)			
q_{pp}	$q_{pt} + q_{pV}$	9.064517961275140E-12	J / C.K
Specific charge capacity for the proton (constant *pressure*)			
Q_t	$e.q_{et}$ $e'.q_{pt}$	6.65753003344558E-09	J/K
Charge capacity (constant *temperature*); equal to **R** & C_t			
Q_V	$e.q_{eV}$ $e'.q_{pV}$	9.98629505016837E-09	J/K
Charge capacity (constant *volume*); equal to C_V			
Q_p	$e.q_{ep}$ $e'.q_{pp}$	1.66438250836139E-08	J/K
Charge capacity (constant *pressure*); equal to C_p			
Table B5			

2B.6   Microstates

Symbol	Formula	Value	Units
N_t	$\exp(c_p.L_n(\underline{T}) / R_a)$ $\exp(\mathbf{q_p}.L_n(\underline{T}) / R_a)$ $\exp(2.5 . Ln(\underline{T}))$		
Microstate (constant *temperature*)			
N_V	c_v / R_a $\mathbf{q_v} / R_a$		
Microstate (constant *volume*); N_t multiplied by 1.5			
N_p	c_p / R_a $\mathbf{q_p} / R_a$		
Microstate (constant *pressure*); N_t multiplied by 2.5			
Table B6			

2B.7   Electricity

Apart from the Farad, no values are provided for the following electrical properties because the Amp, Volt, Ohm and Henry are now redundant.

Symbol	Formula	Value	Units
I	$e.f$		C/s
Electrical current (Coulomb flow-rate)			
V	PE/e		J/C
Electrical voltage (potential energy per coulomb)			
Ω	$V/A = PE / f.e^2$		$J.s/C^2$
Electrical resistance (momentum over distance per Coulomb squared)			
H	$\mu_o.R$		$G.m^2/C^2$
Henry; unit of mutual inductance			
F	$e.N_A$	4.53600855374358E-05	C/mol
Farad; unit of electrostatic capacitance (equal to **RACe**)			
P	$V.A = PE.f$		J/s
Power (Watt)			

Table B7

PE is the potential energy between a proton and its orbiting electron

2B.8   Newton vs Planck

Newton's atom is the real one; the one we see, hear and feel all around us. Planck's atom is theoretical, it does not exist. It is based upon his three constants; time (t^P), length (A^P) and mass (m^P). See Table A8 below.

It should be noted here that Planck's length is actually an orbital radius (R) and not a wavelength (λ) as depicted below. The reason being; force (F) is Energy divided by length, and a length in such a formula can only be an orbital radius.

The atomic particles in Planck's atom are identical in size; his electron is exactly the same, in all respects, as his proton. The only consistency between the two atoms (Newton's and Planck's) is that the product of the particle volumes in each atom is identical; $V_e.V_p = V^P.V^P = \Sigma$

It is the comparison between these two atoms that has given us the ability to solve Newton's gravitational constant (G; see Chapter 3.6.3)

The following Tables (B8a to B8c) provide comparative results for three atomic variations in which the particle properties are defined thus:

Newton atom; comprising Newtonian particles (Table B8a)
$t = a_o/c$; $A = a_o$; $m = \rho_u$; $E = \rho_u.c^2$; $F = E/a_o$

Planck^N atom; comprising Newtonian particles (Table B8a)
t^P; A^P; m^P; E^P; F^P (Table B8b)

Planck^P atom; comprising Planck particles (Table B8c), and calculated using a revised version of his own constant; ℏ:
$ℏ = \sqrt{[\pi.m.e^2.a_o / \varepsilon_o]} = 8.61.0229382467$ G.m²/s
in which:
$a_o = \lambda^P / 4\pi^2$
$e = \sqrt{[G.m^{P2} / k.\varphi]}$
$m = m^P$

Note: 'c', 'ε_o', 'G', 'φ' and 'k' are equal in both Newton and Planck atoms

Newton atom compared with a Planck^N atom:

	Newton Atom (A)	Planck^N Atom (B)	A/B
t	2988.7517123993	6.10208816363872	489.7916307091
λ	18778.8808461551	38.3405506928893	489.7916307091
m	1750.60379891304	3.57418071104766	489.7916307091
E	69111.0678332636	141.102998704179	489.7916307091
F	3.68025487777744	3.68025487777744	1
Table B8a			

Newton atom compared with a Planck^P atom:

	Newton Atom (A)	Planck^P Atom (B)	A/B
t	2988.7517123993	1532.50363357728	1.95024119154794
λ	18778.8808461551	9629.00431369208	1.95024119154794
m	1750.60379891304	897.634511310653	1.95024119154794
E	69111.0678332636	35437.1900936053	1.95024119154794
F	3.68025487777744	3.68025487777744	1
Table B8b			

Planck^N atom compared with a Planck^P atom:

	Planck^N Atom (A)	Planck^P Atom (B)	A/B
t	6.10208816363872	1532.50363357728	251.144131726775
λ	38.3405506928893	9629.00431369208	251.144131726775
m	3.57418071104766	897.634511310653	251.144131726775
E	141.102998704179	35437.1900936053	251.144131726775
F	3.68025487777744	3.68025487777744	1
Table B8c			

Note: 489.7916307091 ÷ 1.95024119154794 = 251.144131726775

2B.8.1   Newton's Atom (particle properties)

The following Table provides the properties of particles in the atom that exists in nature.

Symbol	Formula	Value	Units
m_e		1	
electron mass (Table 2B.1)			
m_p		1836.15115053207	
proton mass			
m_n	$m_e + m_p$	1837.15115053207	G
neutron mass			
V_e	m_e / ρ_u	1.27822236702922E-47	m^3
electron volume			
V_p	m_p / ρ_u	2.34700946985653E-44	m^3
proton volume			
V_n	m_n / ρ_u	2.34828769222356E-44	m^3
neutron volume			
r_e	$\sqrt{[\, 3.V_e / 4\pi\,]}$	1.45046059426276E-16	m
electron radius			
r_p	$\sqrt{[\, 3.V_p / 4\pi\,]}$	1.77613270336827E-15	m
proton radius			
r_n	$\sqrt{[\, 3.V_n / 4\pi\,]}$	1.77645508248591E-15	m
neutron radius			
t^N	a_0 / c	2988.7517123993	s
Newton time			
A^N	a_0	18778.8808461551	m
Newton EME amplitude			
m^N	$a_0.c^2 / G$	7.82336489952175E+46	G
Newton mass			
E^P	$m^N.c^2$	1.78455755577097E+65	J
Newton energy			
F^N	E^P / A^N	9.50300271028316E+60	N
Newton force			
ρ_u	$m_e . \sqrt{[\xi_m/\Sigma]}$	7.82336489952175E+46	G/m^3
The ultimate density of Newton's atomic particles			
Table B8.1			

2B.8.2   Planck's Atom (particle properties)

The following Table provides the properties of particles in a fictitious atom that has been constructed using Planck's values; t^P, A^P, m^P

Symbol	Formula	Value	Units
m_e^P	m^P	2.1765500017459E-08	kg
electron mass			
m_p^P	m^P	2.1765500017459E-08	kg
proton mass			
m_n^P	N/A	N/A	N/A
neutron mass			
V_e^P	$\sqrt{\Sigma}$	5.47722557505166E-46	m^3
electron volume			
V_p^P	$\sqrt{\Sigma}$	5.47722557505166E-46	m^3
proton volume			
V_n^P	N/A	N/A	N/A
neutron volume			
r_e^P	$\sqrt{[\ 3.V_e^P\ /\ 4\pi\ ]}$	5.07563837996471E-16	m
electron radius			
r_p^P	$\sqrt{[\ 3.V_p^P\ /\ 4\pi\ ]}$	5.07563837996471E-16	m
proton radius			
r_n^P	N/A	N/A	N/A
neutron radius			
t^P	$\sqrt{[\ \hbar.G\ /\ c^5\ ]}$	5.39096122598359E-44	s
Planck time			
A^P	$\sqrt{[\ \hbar.G\ /\ c^3\ ]}$	1.61616952231128E-35	m
Planck EME amplitude			
m^P	$\sqrt{[\ \hbar.c\ /\ G\ ]}$	2.1765500017459E-08	kg
Planck mass			
E^P	$m^P.c^2$	1.95618559889902E+09	J
Planck energy			
F^P	$c^4\ /\ G$	1.21038391820525E+44	N
Planck force			
ρ_u^P	$m^P\ /\ V_e^P$	3.97381844498046E+37	kg/m^3
The ultimate density of Planck's atomic particles			
Table B8.2			

3   Physical Properties

The following chapters describe the atomic & astronomic physical properties and what they actually mean.

The calculations and conversions provided in this chapter are based upon Chapter 2; **Part A**.

3.1 Gravity

Gravity is the potential energy between all *masses* due to the non-polar magnetic charge in their atomic particles (chapter 3.2).

That '*gravity is magnetism*' changes nothing in terms of force and energy in the laws of motion. Isaac Newton remains correct. Identical results may be achieved either by using *gravity* as explained by Isaac Newton, or by using magnetism as described by William Gilbert and Hendrik Lorentz. However, whilst we know exactly what magnetism is, and can explain it in terms of energy, not even the great man himself knew what *gravity* is and how it works.

All matter comprises atomic particles, each of which carries an *electrical* charge ($\pm e$, $\pm e'$, 0) and a *magnetic* charge (m_e, m_p, m_n).

Lorentz' formula for magnetic attractive force may be used to explain potential energy in terms of *non-polar* magnetism:

$F = q.v.B = q.v/RC$
Note: B = 1/RC (chapter 3.6.16)

Where: 'q' is charge, 'v' is relative velocity, and 'B' is the magnetic field. However, when the attracting *masses* are stationary, the relative velocity in this formula must be modified to use the potential acceleration and radial separation between the two *masses*.

Lorentz used electrical charge 'q' in his formula, but it actually applies to magnetic charge 'm'. To explain:
Any two attracting masses (m_1 & m_2) may be described in terms of their magnetic charges (m_1 & m_2), which, in the case of an atom, are numerically equal to the elementary charge unit (e) (chapter 3.6.7). Because we already know that every particle possesses a known charge, we can determine the total charge (q) in any mass (m) as follows:
$q = m.e/m_e$
where: m_e is the mass of an electron

Lorentz' magnetic force formula; $F = q.v/RC$ is not very helpful for calculating the magnetic force between stationary particles but because;

$a = v^2/R$ …

… we can transform Lorentz' force formula for potential (rather than kinetic) force (F) and energy (E) as follows:
$F = q.a/RC$ (kg.m/s²) & $E = q.a.R/RC$ (kg.m²/s²)

Which can be corroborated with Newton's and Coulomb's formulas for the magnetic potential energy between a proton and its orbiting electron:
$F = PE/R$ {N} & $PE = e.v^2/RC$ {J}

All of which is simply Lorentz's formula in a format suitable for stationary *masses*.

51

Moreover; **Exactly** the same potential energy (PE) between a proton and its orbiting electron can be found using any and all of the following formulas:

Category	Formula	Calculation Result	Units
Orbital:	$PE = m_e.a.R$ *(Newton)*	3.94042969432572E-20	J
Potential:	$PE = G/\varphi . m_e.m_p/R$ *(Newton)*	3.94042969432572E-20	J
Electrical:	$PE = k.e^2/R$ *(Coulomb)*	3.94042969432572E-20	J
Magnetic:	$PE = e/RC . g.R$ *(Lorentz & Dixon-Roche)*	3.94042969432572E-20	J
Heat:	$PE = T.k_B.Y$ *(Dixon-Roche)*	3.94042969432572E-20	J

Table 2.2-1 Energy Calculation Methods ($\underline{T}$ @ 300K)

$v = 207982.67075397$ m/s *(electron velocity)*
$a = 7.38815108322488E{+}18$ m/s^2 *(potential acceleration between m_p and m_e)*
$R = 5.85488721693451E{-}09$ m *(electron orbital radius)*
Where:
electron; $m_e = 9.1093897E{-}31$ {kg}
proton; $m_p = 1.67262163783E{-}27$ {kg}
$e = 1.60217648753E{-}19$ {C}
$RC = 1.75881869180545E{+}11$ {C/kg}

The two important constants here are Isaac Newton's *gravitational* constant (G) and Coulomb's constant (k), both of which are based upon the properties of Quanta. It is important, therefore, to establish their relationship to each other.

Because *exactly* the same forces and energies can be obtained using Lorentz's, Coulomb's and Newton's force formulas in both celestial and atomic matter, it is clear that *gravity* is magnetism.

Therefore, we can apply the same argument to celestial bodies:

Coulomb's constant:
$k = R_n.m_e.c^2/e^2 = 8.98755184732667E{+}09$ {kg.m^3 / s^2.C^2}

Newton's *gravitational* constant:
$G = a_o.c^2 / m_u = 6.67359232004332E{-}11$ {m^3 / kg.s^2 per m^3}
$\quad = k.e^2.\varphi / m_e.m_p = 6.67359232004332E{-}11$ {m^3 / kg.s^2}
$F = G.m_1.m_2/R^2$ **#1**

After dividing out the *mass* components:
$m_1.m_2 \div m_e.m_p$, we get a product of particles '$n_1.n_2$'

Let $G_e = k.e^2.\varphi = 1.01682605280249E{-}67$ kg.m^3 / s^2

And rewriting Newton's force formula thus: $F = G_e.n_1.n_2 / R^2$
we get identical results to **#1** above, but it is now in terms of charge units.

This calculation for the electrical potential energy between the sun and the earth at its perigee, gives: PE = 1.2208949335E+73 J, whereas the actual potential energy is: PE = 5.380981972219E+33 J, the difference between the two is the coupling ratio; 'φ': magnetic force ÷ electrical force!

Whilst magnetic energy is accrued, electrical energy is shared, locking the electrical attractive energy between an electron and its proton at the atomic level. This [electrical] energy does not pass beyond the atom. I.e. 'G_e' may only be used instead of 'G' in Isaac newton's formula for the atom. 'G_e' cannot be used for the calculation of lunar, solar or galactic orbital systems; conclusively demonstrating that Isaac Newton's laws of orbital motion apply equally well to atoms.

Also, between atoms (Z = atomic number) …

magnetic: $a = G/\varphi \cdot Z.m_p/d^2$

electrical: $a = k/m_e \cdot Z.(e/d)^2$

… both of which give *exactly* the same results, confirming:

a) magnetism is accrued, and;

b) electricity is shared, and;

c) the Newton-Coulomb atomic model.

This discovery unites Newton's *gravitational* laws of orbital motion with those for the atom, something that cannot, and no doubt ever will be the case for quantum theory.

It is no longer necessary to use the term *gravity*; **Gravity is Magnetism**.

3.2 Mass

The term *'mass'* is an unknown concept. It is today used to describe the inertia of matter; i.e. its resistance to movement. This resistance is actually the reciprocity between a body's magnetic charge and the magnetic field in its universal environment.

I.e. *mass* (m) is magnetic charge (m), the magnitude of which is the non-polar equivalent of the elementary charge unit: $m \equiv |e|$
Therefore; the elementary charge unit ($\pm$e) should be referred to as the *electrical charge unit* (e), and *mass* should be referred to as the *magnetic charge unit* (m). The non-polar nature of the magnetic charge is what generates attraction between all particles.

Because the magnetic charge (m) in Quanta is constant, the magnetic fields they generate (μ) will also be constant; just as *mass* is constant.

We currently refer to the magnetic field generated by this magnetic charge as *gravity* (chapter 3.1), which is dependent upon the radial distance(s) between two (or more) bodies.

The electron always holds a constant magnetic charge of 'm_e' and a constant electrical charge '-e'.

The proton always holds a constant magnetic charge of '$\xi_m.m_e$' …
      … and a constant electrical charge '+e' whilst it <u>does not</u> host an electron partner.
      … and an electrical charge that varies between '+e' and '+e.ξ_v' whilst it hosts an orbiting electron partner.

Example calculations using *magnetic charge* are provided below, in which the magnetic charge (m) of an electron is given the same numerical value as the elementary *electrical* charge unit; 'e' (Coulomb)

$$\gamma = e/m_e = 1.75881869180545E+11 \ G/kg$$

Where: 'γ' is the factor used to numerically convert *mass* to the elementary *magnetic* charge (m). I have chosen the unit-name '*Gilbert*', with the symbol; '*G*', in deference to William Gilbert:

I.e. $m_e = 1.60217648753E-19 \ G$ and $1kg = 1.75881869180545E+11 \ G$

Isaac Newton's calculations were based upon *mass*, as was his *gravitational* constant:
      $G = a_o.c^2/m_u$ (Table 2a)
      $G = 6.67359232004333E-11 \ m^3 / kg.s^2$
and his force formula: $F = G.m_1.m_2/R^2$

We can revise the above for magnetic charge thus: $M = a_o.c^2 / m_u.\gamma^2$
      $M = 2.15733469430661E-33 \ m^3 / G.s^2$

Newton's (and Gilbert's) force formula still applies, but; m_1 & m_2 are now the magnetic charges of each body in *Gilberts* (*mass* multiplied by 'γ').

We currently calculate the Coupling Ratio (φ) using *mass* & 'G' as follows;

$G = a_o.c^2/m_N = 6.67359232004333E-11$ m^3 / s^2.kg

$k = R_n.c^2.m_e/e^2 = 8.98755184732666E+09$ kg.m^3 / C^2.s^2

$\varphi = G.m_e.m_p / k.e^2 = \xi_m.G/k . (m_e/e)^2 = \textbf{4.40742111792334E-40}$

but if we set the magnetic charge of an electron (m_e) to the same value in Gilberts as 'e';

$\gamma = e/m_e$

$M = a_o.c^2/(m_N.\gamma) = 3.79436058482686E-22$ m^3 / s^2.G

$k = R_n.c^2.(m_e.\gamma)/e^2 = 1.58074741826487E+21$ G.m^3 / C^2.s^2

$\varphi = G.m_e.m_p / k.e^2 = \xi_m.G/k = \textbf{4.40742111792334E-40}$

We currently calculate the potential energy in the earth at its orbital perigee thus:

$PE = G.m_1.m_2/R = \textbf{-5.38099811251204E+33}$ J

Alternatively, we could calculate the potential energy in the earth at its orbital perigee using magnetic charge & 'M' as follows:

Magnetic charge in our sun:

$m_1 = 1.9885E+30$ x $1.75881869180545E+11$
$\quad = 3.4974109686551E+41$ G

Magnetic charge in the earth:

$m_2 = 5.96451976771313E+24$ x $1.75881869180545E+11$
$\quad = 1.0490508855097E+36$ G

$R = 147095000000$ m

$PE_G = -M.m_1.m_2 / R = -9.46395719566978E+44$ G.m^2/s^2

$PE_N = \gamma.PE_G = \textbf{- 5.3808600282471E+33}$ kg.m^2/s^2 (J)

If we let 1 *Gil* = 1E+09 Gilberts, then;

1g = 175.881869180545 *Gil*

1kg = 0.175881869180545 *Gil*

3.3 Electricity

The operational properties that define electrical performance through elemental matter, are power, Voltage and resistivity; current and resistance are consequences.

The fundamental electrical relationships are:

Voltage: $V = I.\Omega$ {J/C}

current: $I = V/\Omega$ {C/s}

power $= V.I$ {J/s}

resistance: $\Omega = V/I$ {J.s/C^2}
resistivity: $\rho = \Omega.A/\ell$ {J.s.m/C^2}

Natural DC: The electricity within a proton-electron pair is calculated thus:

$V = PE/e = m_e.v^2 / e = m_e.T_N / X.e$ {J/C}

$I = e.f = e . (T_N/T_n)^{1.5} / t_n$ {C/s}

power $= V.I$ {J/s}

$\Omega = V/I$ {J.s/C^2}

Unnatural DC: The electricity in elemental matter is calculated thus:

$P =$ input value {J/s}

$V =$ input value {J/C}

atomic $V_a = {}_{n=1}\Sigma^{n=Z} PE_n / e$ {J/C}
the conductor's highest operating (core) temperature is: $T . V/V_a$

$I = P/V$ {C/s}

$\Omega = V/I$ {J.s/C^2}

atomic $\Omega_a = PE / f.e^2$ {J.s/C^2}

Resistivity: Electrical resistivity in a conductor is generally calculated thus:

$\rho = \Omega.A/\ell$ {J.s.m/C^2}

where: A is its cross-sectional area and ℓ is its length

The electrical resistivity of elemental matter, which varies with temperature, is based upon its atomic spacing $(d = {}^3\sqrt{[m_a/\rho_m]})$;

resistivity: $\rho = \Omega_a .d^2/2\pi R$ {J.s.m/C^2}

3.4 Magnetism

Joseph Henry's magnetic constant (μ_o) gives us the minimum magnetic field; at the neutronic condition:

$$\mu_o = m_e.R_n/e^2 \quad \{kg.m/C^2\}$$

and magnetic force within a proton-electron pair at any temperature (R & f) can be calculated thus ...

$$F_M = \mu.I^2 . (2\pi)^2 = m_e.v^2/R \qquad \{kg.m/C^2 \text{ x } (C/s)^2 = kg.m/s^2\}$$

where $\mu = m_e.R/e^2$ & $I = e.f$

The reason magnetism does not vary with temperature:

The magnetic potential force in a proton-electron pair is defined by Joseph Henry:

$$F_M = \mu.I^2 . (2\pi)^2 \qquad \{kg.m/s^2\}$$

$$\mu.I^2 = (m_e.R/e^2) \text{ x } (e.2\pi f)^2 \text{ x } (4\pi.R^2 / 4\pi.d^2)$$

remove the constants at any distance 'd' and we get:

$$\text{factor} = R.f^2.R^2 = R^3.f^2 = 6.41524280848628 \ \{m^3/s^2\}$$

the reciprocal of Isaac Newton's constant of proportionality for the proton-electron pair:

$$K = t_n^2/R_n^3 = 0.15587874533403 \ s^2/m^3 = 1/\text{factor}$$

i.e. a constant!

The PHYSICAL CONSTANTS

3.5 Heat, Light & Temperature

Heat is what we feel from the electro-magnetic energy (EME) emitted by proton-electron pairs. Temperature is how we measure this heat. Each atomic shell will emit heat at a different temperature, the highest being that emitted from the innermost shell and that which we measure.

Heat is the EME radiated by a proton-electron pair: the greater the energy, the greater the heat.

Only proton-electron pairs radiate EME; lone protons and electrons can't.

All atoms comprise proton-electron pairs. The lower the atomic shell number in which an electron partner is orbiting, the higher the EME emitted by the proton-electron pair.

For example:

More EME will be emitted by an atom with a neutral electrical charge at a given temperature, than with a positive ion of the same atom at the same temperature.

Also, much more EME will be emitted by a kilogram of boron at a given temperature than by a kilogram of uranium at the same temperature. This is because far more of the proton-electron pairs in boron are in lower shell numbers.

All heat energy is radiated:

Convection and conduction are simply different forms of radiation.

Conduction is the transfer of EME radiation between proton-electron pairs within viscous matter.

Convection is the balancing of electrical forces between adjacent atoms. Atoms at high temperature (with greater heat energy) will try to move to a position where electrical repulsion between the nucleic protons of adjacent atoms can balance. In other words, hot atoms in an unconstrained container (e.g. an atmosphere) will try to move to where atmospheric density is least; i.e. where gaseous density is lowest. In an atmosphere, this is always *upwards*, away from the earth's surface.

3.5.1   Heat

Heat is the EME generated by the kinetic energy in an atom's electrons. The heat we *feel* is from the senses we have developed to tell us when this kinetic energy is too high or too low. Electron kinetic energy is generated by the EME it absorbs from its surroundings.

It is important to remember that all the EME generated in the universe is just that; EME. It possesses; no light or heat – nothing, apart from energy, and the entire spectral band is simply a single range of EME from 1.692E+22Hz to 3000Hz. We [humans] have split the spectrum into special bands; "γ, X, ultra-violet, light, infra-red, micro, radio" for our own convenience, these bands will mean nothing to any other form of life.

If you or I, devoid of electrons – impossible I know, but bear with me – were to sit in the space between the sun and the earth, we would not be able to detect the sun's radiated EME. It would be invisible in every sense to the fictitious you (or me). EME is useless to all forms of life unless it can be detected.

Whilst EME doesn't deteriorate with distance travelled, we don't feel the sun's surface temperature (5778K) here on Earth because the energy *density* (Joules per square metre) radiated at the sun's surface is distributed over a spherical surface area between 45,000 and 48,000 times greater (dependent on the time of year). Therefore, the EME *density* we receive will be correspondingly less.

Life here on Earth, has evolved to detect and use this energy through our complex molecules. The trouble is, such molecules have energy tolerance levels, outside which they would no longer function; i.e. their state-of-matter, strength or condition (gas-viscous) could change, or inter-atomic bonding could fail.

If a block of viscous iron, the function of which is to be solid, received sufficient EME to increase its proton electrical charge energy above that of its atomic magnetic field energy, it would become a gas. And it would cease to be *a block of iron*; i.e. no longer functional.

You can't damage a block of iron, so it doesn't need senses. It doesn't matter how many times you change it from gas to viscous and back again it always returns to iron. The higher its temperature the stronger its atoms remain, until the innermost electrons achieve the *speed of light*, when it will become a different element (Z-1 or Z-2).

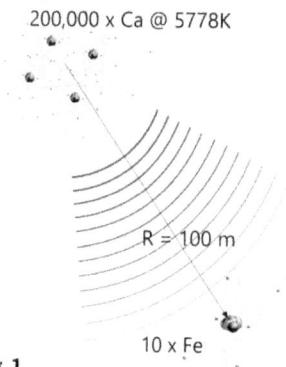

200,000 x Ca @ 5778K

R = 100 m

10 x Fe

Fig 1

We (humans) have five senses – if you exclude time – smell, touch, taste, sight and hearing; each of which were developed for use and protection.

There are tolerance levels regarding acceptable amounts of EME any living organism can receive and remain functional. Therefore, all living organisms have developed senses, that can be used to ensure that these tolerance levels are maintained.

All the EME in our environment is shared between all of our electrons. The greater the EME *density*, the greater the *heat* we feel. Irrespective of the *temperature* of the atoms that generated the EME, if the energy *density* can be shared throughout all the electrons in our body without exceeding its tolerance levels, we will remain functional. For example (Fig 1):

200,000 calcium atoms at 5778K will radiate 4.446E-13 J of EME.

A body of 10 iron atoms with a surface area 100th that of the calcium atoms, 100 metres away would absorb a total of 3.54E-20 J of heat energy, resulting in a temperature rise of 9K in the iron.

In other words; it is not the *temperature* of the atoms emitting the EME that defines our body-temperature; it is the quantity of EME absorbed by our body's electrons.

3.5.2   Light

What we humans have designated as light is EME with a wavelength range between 4E-07m (blue) to 8E-07m (red). We (humans) only call it light because we can *see* it with an unaided eye.

Ultra-violet and infra-red are also EME ranges, but we don't call them light because we can't *see* such wavelengths without artificial aids.

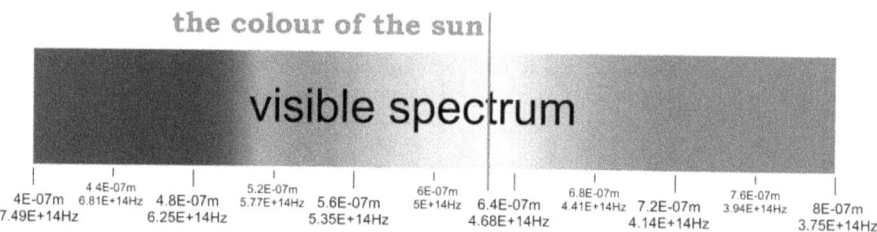

the colour of the sun

visible spectrum

| 4E-07m | 4.4E-07m | 4.8E-07m | 5.2E-07m | 5.6E-07m | 6E-07m | 6.4E-07m | 6.8E-07m | 7.2E-07m | 7.6E-07m | 8E-07m |
| 7.49E+14Hz | 6.81E+14Hz | 6.25E+14Hz | 5.77E+14Hz | 5.35E+14Hz | 5E+14Hz | 4.68E+14Hz | 4.41E+14Hz | 4.14E+14Hz | 3.94E+14Hz | 3.75E+14Hz |

Fig 2

Question: *"If colour is defined by EME, and the surface of the sun is at a temperature of 5778K and looks yellow; how can my towel (at 300K) also look yellow?"*

Colour is a range of EME wavelengths that we cannot detect until it is absorbed by the electrons in our optical receptors (eyes).
Light is the intensity of visible EME wavelengths, or put simply; the number of EME rays per square metre.

We (humans) have developed optical sensors (eyes) to detect a particular bandwidth of EME that best suits our purpose. Other lifeforms have developed receptors that best suit their own environment that may be outside the aforementioned *optical* range. Irrespective of a lifeform's preferred optical bandwidth, the purpose of sight is the same; to see what's in its environment and how best to exploit it.

Every proton-electron pair in the universe, *including that block of wood in the garden*, emits EME at a wavelength commensurate with the kinetic energy in its electrons. But you cannot see the EME radiated by that block of wood here on Earth, because it is radiated in the infra-red range. This is the reason infra-red cameras reveal objects that naturally emit EME in the dark here on Earth (e.g. living organisms). They are actually capturing the electro-magnetic energy given off by the objects themselves, not the electro-magnetic energy radiated by the sun.

Unlike in a prism, the diffraction of light through natural matter is not organised. The image of that block of wood, is the sun's EME reflected and/or refracted by or through its constituent atoms and molecules.

Just as with heat, if the EME received by your eyes is not too intense, i.e. its density remains within your body's tolerance levels, the sun's rays you see will do you no harm.

Whilst our sun's *atmosphere* comprises proton-electron pairs as hydrogen and helium due to its heat, its internals comprise all of nature's elements, the proton-electron pairs of which give off EME according to their shell numbers. A star therefore emits a preponderance of electro-magnetic radiation at all the wavelengths from yellow to gamma. All the infra-red, micro-wave and radio-wave energy coming to us (from extra-terrestrial sources) is generated by colder celestial bodies in the universe (e.g. planets or stars with few or no satellites).

So, when you see a *yellow* towel here on earth, your eyes are detecting the EME radiated by the sun at a wavelength of $\approx$6.3E-07m diffracted by the molecules in your towel, but at an intensity that will not harm your eyes (refer to Fig 1).

3.5.3   Temperature

Temperature is not a genuine physical variable, neither is it a measure of the heat emitted by a body of matter; it is simply a convenient term for describing one of life's senses (touch), which can detect the highest electron kinetic energies in our body's atoms.

$$T = PE / Y.k_B$$

Temperature we measure is the electro-magnetic energy (heat) emitted by the proton-electron pairs whose electrons are orbiting in an atom's innermost electron shell (shell-1); the highest EME emitted by an atom.

$$T_1 = PE_1 / Y.k_B$$

The highest temperature that can be generated by a proton-electron pair occurs immediately before the two particles unite to become a neutron; i.e. when the electron is orbiting at 'c':

$$T_n = PE_n / Y.k_B = 623316124.717178 \text{ K}$$

Nothing in the universe can exceed this temperature, which therefore represents the highest *possible* naturally generated electro-magnetic energy. This is the temperature at the core of all bright stars.

The lowest possible temperature that can be generated by a proton-electron pair occurs immediately before the electron leaves its proton partner. This is referred to as the 'cold' temperature:

$$T_x = X.(c / Y.\xi_m)^2 = 2.04274907568265 \text{ K}$$

Below this temperature, all matter will exist as lone particles (protons and electrons), emitting no EME. There is no such thing as zero temperature because $T = X^R/R$. At zero temperature, the electron's orbital radius would have to be infinite. The electron would have left its proton partner long before this occurred.

The temperature of outer space is approximately 2.7255K. The WMAP, & COBE projects were launched to measure and map universal temperature that was claimed to be left over from the last 'Big-Bang'. However, because EME travels at a constant speed of 'c', and the matter ejected from the last 'Big-Bang' was initially travelling at 0.6% 'c', and is today travelling at just 0.07% 'c', the heat in outer-space cannot possibly left over from the last 'Big-Bang'. It can only be that generated and radiated by all the active celestial bodies in the universe.

It is important to understand that there are two principal means of measuring EME energy; Boltzmann and SHC, but they appear to generate different temperatures, for example, a proton-electron pair @ 300K;

Boltzmann: $E_B = k_B.Y.T = 3.94042969432577E-20 \text{ J}$
SHC: $E_S = SHC.Y.m_m.T = 1.97021484716289E-20 \text{ J}$
$E_B = 2.E_S!$

However, EME has two measurable energies; *magnitude* (KE) and *range* (PE).

Magnitude is a measure of the *kinetic* energy in the orbiting electron generating the EME, and is used in the calculation of specific heat capacity; SHC = J / K.kg, where 'J' refers to EME magnitude.

Range is a measure of the energy 'span' in EME that defines electro-magnetic energy using Boltzmann's constant (k_B): $|KE^-| + |KE^+| = 2.KE = PE$
which is equal to the *potential* energy in the proton-electron pair generating the EME.

When quoting EME energy generated, it is usual to specify its *magnitude*, rather than its *range*, but they both equally apply.

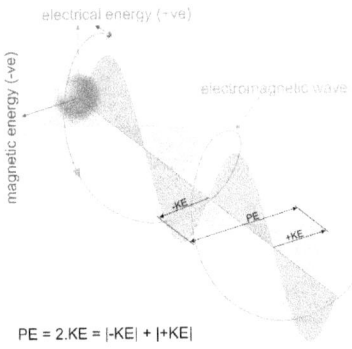

PE = 2.KE = |-KE| + |+KE|

Therefore, both calculations are correct because they give the same temperature:

$$T = KE / SHC.Y.m_n = PE / Y.k_B$$

Where:
KE = the orbiting electron's kinetic energy
SHC = the specific heat capacity of the proton-electron pair (a hydrogen atom)
Y = temperature constant
m_n = the mass of the pair ($m_e + m_p$)
PE = the potential energy in the pair (PE = 2.KE)
k_B = Boltzmann's constant

3.5.4   PVRT

Confirmation of this atomic model can be deduced from the ease with which the well-known and accepted gas-pressure calculation formula:

$$p = R_i/(RAM/1000) \cdot T.\rho = R_a.T.\rho$$

can be replaced with a formula using Newton's and Coulomb's formulas thus;

$$p = PE_1/Y \cdot \rho/m_a$$

where; PE_1 is the potential energy in the atom's proton-electron pair(s) in shell-1, ρ is the gas density and m_a is the atomic mass

The following Table shows a calculation comparison for our atmospheric gases @ 300K, using three different calculation methods, confirming that the Newton-Coulomb atomic model is valid:

$$p_1 = R_a.T.\rho \quad p_2 = PE_1/Y.d^3 \quad p_3 = k_B.T/d^3 \quad p_4 = k_B.\rho.T/m_a \quad p_5 = PE_n/Y \cdot T/T_n/d^3 \quad N/m^2$$

	Helium	Nitrogen	Oxygen	Fluorine	Neon	Chlorine	Argon
ρ	9.17599E-07	0.964387867	0.295545871	1.60E-05	1.60E-05	0.003214	0.016379
p_1	0.567390018	170406.8635	45716.82756	2.084366376	1.964042434	224.3696207	1014.760406
p_2	0.567390018	170406.8635	45716.82756	2.084366376	1.964042434	224.3696207	1014.760406
p_3	0.567390018	170406.8635	45716.82756	2.084366376	1.964042434	224.3696207	1014.760406
p_4	0.567390018	170406.8635	45716.82756	2.084366376	1.964042434	224.3696207	1014.760406
p_5	0.567390018	170406.8635	45716.82756	2.084366376	1.964042434	224.3696207	1014.760406

3.6 Important Constants (explained)

It is a fact that many natural constants are incorrect, approximate or misunderstood. It is the aim of this Chapter to finally resolve this issue.

Each of the following sub-Chapters describe the reasoning behind the units and values given for the most important physical constants; where they have come from and how they have evolved.

3.6.1   Σ

This universal constant is the reciprocal of the product of atomic particles (protons and electrons) in a unit mass of ultimate density:

ultimate density: $\rho_u = m_e \div (4/3\pi.r^3) = 7.1266079635045E+16$ kg/m³
Newtonian mass: $m_u = 7.1266079635045E+16$ kg
Number of electrons: $\epsilon_e = m_u / m_e$
Number of protons: $\epsilon_p = m_u / m_p$
$\Sigma = 1 / (\epsilon_e.\epsilon_p) = 3E-91$

Whilst the reasoning behind this bizarrely accurate value has yet to be discovered, it applies to all the following relationships, including the unification of Newton's and Plack's particles:

$G = \sqrt{[\ \Sigma.a_o^2.c^4 / m_p.m_e\ ]}$

$m_e.m_p = \Sigma.\rho_u^2$

$\rho_u = m_e.\sqrt{[\xi_m/\Sigma]}$

$V_e.V_p = \Sigma$
$V^{P2} = r^{P6}.(4/3.\pi)^2 = \Sigma$

$r^P = \sqrt[6]{[\Sigma / (4/3.\pi)^2]} = 5.07563837996471E-16$ m
is the radius of a Planck particle

$\xi_m = (\varphi.R_n / a_o)^2 / \Sigma$

$\xi_m = \Sigma.(\rho_u/e)^2$

$\xi_m = \varphi^2 . (4\pi/\xi_v)^4 / \Sigma$

$m_u/m_e . a_o/R_n = \varphi/\Sigma$

If $F^N = G.m_e.m_p / a_o^2$ then; $F^N/F^P = \Sigma$
(Table A8; Tables A to C; F^P)

Its units *appear* to be m⁶. I say *"appears"* because it also unites the Newton and Planck forces (F^N & F^P and φ/Σ; see above), which means it can also have no units.

The PHYSICAL CONSTANTS

Σ defines the ratio (ϑ);
"electron orbital radius" : "radial separation (centres)"

That is; the relative radii of the orbiting electron and the "electron radius plus proton radius" ($ϑ = R /$ $[r_e+r_p]$)

For example; if Σ = 3E-91, at a velocity of 'c'; $ϑ = 1.46677550700175$

The following Table shows the changes that would result in modifying Σ to give us $ϑ = 1.0$ (i.e. the particles touch at 'c').
All other physical constants unchanged.

Σ		3E-91	2.98746723133494E-90	{m⁶}
ϑ		1.46677550700175	1.0	
G		6.67359232004334E-11	2.10596243650527E-10	m³ / kg.s²
ρ_u		7.12660796350450E+16	2.25835348953929E+16	kg/m³
φ		4.407421117923340E-40	1.39083463164654E-39	
r_e		1.45046059426276E-16	2.12750007353581E-16	m
r_p		1.77613270336827E-15	2.60518794648537E-15	m

Given that this value (3E-91) for Σ defines; ultimate density, Newton's *gravitational* constant (G) and the neutronic radius (R_n), I have concluded that it must be definitive.

3.6.2 ρ_u

The ultimate (limiting) density is the maximum possible density that exists in nature, and it applies to both atomic particles; the electron and the proton.

$\rho_u = 7.12660796350449E+16 \text{ kg/m}^3$

It may be calculated as follows:

$\rho_u = \sqrt{[m_e.m_p]} / V_e^P$
Refer to Chapter 2A.8.2 for V_e^P

or

$\rho_u = a_o.c^2 / G$
Refer to Chapter 2A.8.1

or

$\rho_u = \sqrt{[m_e.m_p / \Sigma]}$

or

$\rho_u = \epsilon_e.m_e = \epsilon_p.m_p$

3.6.3   G

Today, you will see the units for this constant written as: N.m²/kg², which were units of convenience originally assigned to reflect Newton's formula:

$$F = G.m_1.m_2 / R^2$$

This was because its origin, formula and value were unknown, hence its units were unknown.

Planck gave us the means to determine the solution to Newton's gravitational constant 'G' using his formula for 'length' (which is actually Rydberg's orbital radius and EME amplitude 'a$_o$').

Its Units:

$$\lambda = \sqrt{[G.\hbar / c^3]}$$
$$G = \lambda^2.c^3 / \hbar \quad \mathbf{m^3 / s^2.kg}$$

Its Value:

$$\lambda = \sqrt{[G.\hbar / c^3]}$$

Where: $\hbar = h / 2\pi$ & $\lambda^P = a_o = 5.29177210670E\text{-}11$ m

$$G = \lambda^2.c^3 / \hbar = \mathbf{6.67359232004334E\text{-}11}$$

Its formula:

$$V_e = 4\pi.\lambda^{P2} / R_n.a_o.\xi_v \, (.\, 1 \, m^3)$$
$$\hbar = \tfrac{1}{2}.R_n.m_e.c.\xi_v / 2\pi = R_n.m_e.c.\xi_v / 4\pi$$
$$1/\hbar = 4\pi / R_n.m_e.c.\xi_v = V_e . a_o / 1.m_e.c.\lambda^{P2}$$

Given that; $\lambda^P = \sqrt{[G.\hbar / c^3]}$

$$G = c^3 . \lambda^{P2} . 1/\hbar$$
$$G = c^3 . \lambda^{P2} . V_e . a_o / 1.m_e.c.r^{P2} = c^3/c . a_o . \lambda^{P2}/\lambda^{P2} . [V_e / 1.m_e]$$
$$G = c^2 . a_o \, [V_e / 1.m_e] \, \& \, [\rho_u = V_e / m_e] \, \& \, [m_u = 1m^3.\rho_u]$$
$$\mathbf{G = c^2.a_o / m_u}$$

confirming that G is a constant, not a factor.

Newton's and Kepler's formulas for the constant of proportionality (K) together gave us the correct units for Newton's gravitational constant (G) 300-years ago;

$$K = (2\pi)^2 / G.m = t^2/a^3$$
$$G = (2\pi)^2.a^3 / t^2.m \qquad m^3 / s^2.kg$$

Today we have accurate values for Earth's; orbital axes 'a', orbital period (t), and the mass of its force-centre (the sun), so we also have an accurate value for 'G';

$$(2\pi)^2 \text{ x } 1.49594598077542E\text{+}11^3 \div (1.9885E\text{+}30 \text{ x } 31558118.40^2)$$
$$= \mathbf{6.67359232004334E\text{-}11} \qquad m^3 / s^2.kg$$

And by a remarkable coincidence:

$$c^2.a_o / m_u = \mathbf{6.67359232004334E\text{-}11} \quad m^3 / s^2.kg$$

confirming that gravity is not based upon the mass of matter, it is based upon the particles of which it comprises.

Conclusion

Despite all of which;

why is it still necessary to fabricate its units (N.m²/kg²) from Newton's force-law formula?

why is it still necessary to estimate its value?

why has nobody applied his laws to the atom? because his gravitational constant applies to it.

3.6.4   φ

The coupling ratio is the ratio of magnetic charge energy (E_m) to electrical charge energy (E_e) between a lone proton and a lone electron. It is defined as follows:

$$E_m = G.m_1.m_2 / R \quad \{J\}$$

$$E_e = k.q_1.q_2 / R \quad \{J\}$$

$$\varphi = E_m / E_e = G.m_1.m_2 / k.q_1.q_2$$

$$\varphi = \frac{(6.67359232004334E\text{-}11 \times 9.1093897E\text{-}31 \times 1.67262163783E\text{-}27)}{(8.98755184732667E\text{+}09 \times 1.60217648753E\text{-}19 \times 1.60217648753E\text{-}19)}$$

$$= 4.40742111792334E\text{-}40$$

If magnetism (*gravity*) causes this coupling ratio to be compromised at the core of a body of matter, its neutrons will revert to their base-particles: alpha (proton) & beta (electron). The kinetic energy (in the alpha and beta particles) released by the neutrons will initiate a chain reaction in all neighbouring neutrons breaking the matter apart. This event can only occur naturally in the ultimate-body, where there is sufficient particle *mass* (magnetic charge) to generate the necessary internal pressure and where the atoms are cold.

The following relationships are also true:

$$\varphi = V_p.a_o / R_n$$
refer to Table A8.1 for a definition of V_p

The coupling ratio is the reason we know that matter cannot accrete from hydrogen molecules; the electrical repulsion between proton-electron pairs (hydrogen atoms) is 2.26890050495373E+39 ($1/\varphi$) greater than their gravitational (magnetic) attraction. So, our solar system did not, and in fact could not, have accreted from hydrogen - or any other - gas.

$$\varphi = \gamma/\epsilon_p$$

$$\epsilon_p = m_u/m_p$$

3.6.5   k, k', μ_o, ε_0

Coulomb's constant is equivalent to Isaac Newton's *gravitational* constant (G) when applied to *electrical* force in exactly the same fashion. I.e.:

Newton's formula for potential force: $F = G.m_1.m_2 / R^2$

Coulomb's formula for electrical force: $F = k.q^2 / R^2$

Their quotient (in a proton-electron pair) is the coupling ratio:

$\varphi = G.m_p.m_e / k.e^2 = 4.40742111792335\text{E}-40$

Given the following for the magnetic constant, if:

$\mu_o = 1.0\text{E}-7$ H/m   kg.m/C^2

then;

$\mu_o = R_n.m_e / e^2$ kg.m/C^2

and the unit 'Henry' is actually; kg.m^2/C^2

... and permittivity in a vacuum:

$\varepsilon_0 = 1 / \mu_o.c^2 = e^2 / R_n.m_e.c^2$ C^2.s^2 / kg.m^3 = C^2 / J.m

... electrical energy moment.

Coulomb's constant (k) is normally calculated as follows:

$k = 1/\varepsilon_0 = \mu_o.c^2 = 8.98755184732667\text{E}+09$

but now it may be calculated thus:

$k = R_n.m_e.(c/e)^2 = 8.98755184732667\text{E}+09$ kg.m^3 / C^2.s^2 J.m/C^2

The problem with Coulomb's constant is that it has been allocated units of convenience, making it inconsistent with Newton's gravitational constant 'G' {m^3 / s^2.kg}, but this can be rectified by multiplying Coulomb's constant by the relative charge capacity squared:

$k' = \mu_o.c^2 . RC^2 = 2.78024810626745\text{E}+32$ m^3 / kg.s^2

but because we only know how to calculate force and energy using magnetic charge (kg), we must convert electrical charge to magnetic charge; Coulomb's force formula becomes:

$F_E = k'.(e / RC.R)^2$

$F_M:F_E = G.m_e.m_p / k'.(e/RC)^2 = \varphi$ (the coupling ratio)

Which is correct in both value and units, moreover the ratio of constants is now correct:

$\varphi = \xi_m.G/k'$

3.6.6   h, h'

Because we know that energy cannot be created, it can only be transferred, the electro-magnetic energy emitted by a proton-electron pair must be the same as the kinetic energy in the orbiting electron. Moreover, the frequency (f) and amplitude (A) of electro-magnetic radiation is also equal to that of the orbiting electron (A = R and $f = v / 2\pi R$).

Max Planck claimed that the energy of electro-magnetic radiation can be calculated using his constant as follows:

$$E^P = h.f$$

in which his constant (h) is defined thus:

$$h = \sqrt{[\pi.m_e.e^2.a_o / \varepsilon_o]} = 6.62607174469163E\text{-}34 \textbf{ J.s}$$

However, these units (J.s) can only be correct if a frequency ratio of `1` is applied to his constant thus:

$$h = \sqrt{[\pi.m_e.e^2.a_o / \varepsilon_o]} . f_1/f_2$$

Therefore, the units for Planck's constant should be:

$$\sqrt{[kg.C^2.m / (C^2/m/(kg.m^2/s^2))]} = \sqrt{[kg^2.m^4/s^2]} = \textbf{kg.m}^2\textbf{/s}$$

$$h = \sqrt{[\pi.m_e.e^2.a_o.4\pi.m_e.R_n.c^2 / e^2]}$$

$$= \sqrt{[4\pi^2.m_e^2.c^2.a_o.R_n]}$$

$$\textbf{h} = \sqrt{[(4\pi)^2.a_o . R_n]} . \textbf{½m}_e\textbf{.c} \text{ \{identical units; kg.m}^2\text{/s\}}$$

Planck therefore *actually* identified a range of orbital radii:

a maximum: $R_o = (4\pi)^2.a_o$ (minimum orbital velocity)

a minimum: R_n (maximum orbital velocity)

and a mean: $R_m = \sqrt{[(4\pi)^2.a_o . R_n]}$

Because we know Planck's mean orbital radius:

$$R_m = 4.852618433622630E\text{-}12 \text{ m}$$

and we also know the mean velocity: $v_m = \sqrt{[X^R / X.R_m]}$

$$v_m = 7224342.80705001 \text{ m/s}$$

we can calculate his minimum electron velocity (v_o) from:

$$v_o = c . \sqrt{[R_n / (4\pi)^2.a_o]} = 174090.866621082 \text{ m/s}$$

Given the issue with Planck's units, we can modify his constant thus:

$$h' = h.v_o = \sqrt{[(4\pi)^2.a_o . R_n]} . \text{½}.m_e . v_o.c$$

Note: $v_o.c = v_m^2$

$$\textbf{h'} = \text{½}.R_n.m_e.c^2 = 1.15353857232684E\text{-}28 \textbf{ kg.m}^2\textbf{/s . m/s} \qquad \text{J.m}$$

Planck's maximum temperature (T_n) is the highest possible in nature. It occurs immediately prior to a proton-electron pair uniting as a neutron.

Planck's minimum temperature (T_o) is that below which a proton's electrical charge is constant (e' = e), and above which e' > e. It is also the gas transition temperature of radon, the largest noble atom.

His mean temperature (T_m) is the mean temperature between T_o and T_n.

Planck:	T_o	T_m	T_n
R (m)	8.35643156381572E-09	4.85261843362268E-12	2.817937953839E-15
v (m/s)	174090.866621084	7224342.80705001	299792459
KE (J)	1.38042005551962E-20	2.37714666443634E-17	4.09355561131261E-14
PE (J)	-2.76084011103925E-20	-4.75429332887267E-17	-8.18711122262522E-14
T (K)	210.193328535837	361962.554671561	623316124.717179

Table 3.6.6-1: Planck's Temperatures

Whilst Max Planck's original formula for electro-magnetic energy ($E^P = h.f$) is incorrect, an alternative approach ($E = h'/A$), using the modified version of his constant (h'), does work.

Another interesting relationship with h': $e^2.k/h' = 2$
$$\{C^2 . kg.m^3 / C^2.s^2 . s^2 / kg.m^3 = no\ units\}$$

Moreover; if we extract $\sqrt{[(4\pi)^2.a_o . R_n]}$ from h and modify it slightly, thus:

$$\sqrt{[(4\pi)^2.a_o / R_n]} = 1722.0458764934$$

we get the dynamic ratio ξ_v

It is also interesting to note that the fine-structure constant is equal to:

$$\alpha = e^2 / 4\pi = h'.\varepsilon_o / 2\pi = 2.0427294212227E-39 \ \{C^2\}$$

Planck's modified constant is related to Coulomb's constant thus:

$$h' = h.v_o = \frac{1}{2}.e^2/\varepsilon_o = \frac{1}{2}.k.e^2$$

$$= 1.15353857232684E-28 \ \{kg.m^3/s^2 = J.m\}$$

So, whilst Max Planck's claim regarding electro-magnetic energy using his constant was erroneous, without his work and proposal(s), my solutions would have been much more difficult to achieve; thank you Max! In fact, Max Planck is the *only* 20th Century scientist without whom, none of these scientific discoveries would have been possible.

T	R (A)	v	PE	E^P (error)	E (error)
6.2332E+08	2.8179E-15	299792459	8.18712E-14	1.12193E-11 (274.1)	4.09356E-14 (1)
3.1166E+08	5.6359E-15	211985280.7	4.09356E-14	3.96662E-12 (193.8)	2.04678E-14 (1)
2.0777E+08	8.4538E-15	173085256.9	2.72904E-14	2.15915E-12 (158.2)	1.36452E-14 (1)
1.5583E+08	1.1272E-14	149896229.5	2.04678E-14	1.40241E-12 (137.0)	1.02339E-14 (1)
1.2466E+08	1.4090E-14	134071263.5	1.63742E-14	1.00348E-12 (122.6)	8.18711E-15 (1)
1.0389E+08	1.6908E-14	122389758.9	1.36452E-14	7.63376E-13 (111.9)	6.82259E-15 (1)
8.9045E+07	1.9726E-14	113310898.8	1.16959E-14	6.05785E-13 (103.6)	5.84794E-15 (1)
7.7915E+07	2.2544E-14	105992640.4	1.02339E-14	4.95828E-13 (96.90)	5.11694E-15 (1)
6.9257E+07	2.5361E-14	99930819.7	9.09680E-15	4.15529E-13 (91.36)	4.54840E-15 (1)
6.2332E+07	2.8179E-14	94802699.6	8.18712E-15	3.54785E-13 (86.67)	4.09356E-15 (1)
5.6665E+07	3.0997E-14	90390827.4	7.44282E-15	3.07522E-13 (82.64)	3.72141E-15 (1)
5.1943E+07	3.3815E-14	86542628.5	6.82260E-15	2.69894E-13 (79.12)	3.41130E-15 (1)
4.7947E+07	3.6633E-14	83147467.9	6.29778E-15	2.39359E-13 (76.01)	3.14889E-15 (1)
4.4523E+07	3.9451E-14	80122904.9	5.84794E-15	2.14177E-13 (73.25)	2.92397E-15 (1)
4.1554E+07	4.2269E-14	77406080.1	5.45808E-15	1.93121E-13 (70.77)	2.72904E-15 (1)
3.8957E+07	4.5087E-14	74948114.7	5.11694E-15	1.75302E-13 (68.52)	2.55847E-15 (1)
3.6666E+07	4.7905E-14	72710351.4	4.81594E-15	1.60064E-13 (66.47)	2.40797E-15 (1)
3.4629E+07	5.0723E-14	70661760.2	4.54840E-15	1.46912E-13 (64.60)	2.27420E-15 (1)
3.2806E+07	5.3541E-14	68777107	4.30900E-15	1.35468E-13 (62.88)	2.15450E-15 (1)
3.1166E+07	5.6359E-14	67035631.7	4.09356E-15	1.25436E-13 (61.28)	2.04678E-15 (1)
2.9682E+07	5.9177E-14	65420077.9	3.89862E-15	1.16583E-13 (59.81)	1.94931E-15 (1)
2.8333E+07	6.1995E-14	63915967	3.72142E-15	1.08726E-13 (58.43)	1.86071E-15 (1)
2.7101E+07	6.4813E-14	62511048.9	3.55962E-15	1.01712E-13 (57.15)	1.77981E-15 (1)
2.5972E+07	6.7631E-14	61194879.4	3.41130E-15	9.54220E-14 (55.94)	1.70565E-15 (1)
2.4933E+07	7.0448E-14	59958491.8	3.27484E-15	8.97544E-14 (54.81)	1.63742E-15 (1)
2.3974E+07	7.3266E-14	58794138.4	3.14888E-15	8.46263E-14 (53.75)	1.57444E-15 (1)
2.3086E+07	7.6084E-14	57695085.6	3.03226E-15	7.99687E-14 (52.75)	1.51613E-15 (1)
2.2261E+07	7.8902E-14	56655449.4	2.92396E-15	7.57231E-14 (51.79)	1.46198E-15 (1)
2.1494E+07	8.1720E-14	55670062.1	2.82314E-15	7.18404E-14 (50.89)	1.41157E-15 (1)
2.0777E+07	8.4538E-14	54734364.1	2.72904E-15	6.82785E-14 (50.04)	1.36452E-15 (1)
2.0107E+07	8.7356E-14	53844315.1	2.64100E-15	6.50014E-14 (49.22)	1.32050E-15 (1)
1.9479E+07	9.0174E-14	52996320.2	2.55848E-15	6.19784E-14 (48.45)	1.27924E-15 (1)
1.8888E+07	9.2992E-14	52187168.5	2.48094E-15	5.91827E-14 (47.71)	1.24047E-15 (1)
1.8333E+07	9.5810E-14	51413982.6	2.40798E-15	5.65910E-14 (47.00)	1.20399E-15 (1)
1.7809E+07	9.8628E-14	50674174.5	2.33918E-15	5.41831E-14 (46.33)	1.16959E-15 (1)
1.7314E+07	1.0145E-13	49965409.8	2.27420E-15	5.19412E-14 (45.68)	1.13710E-15 (1)
1.6846E+07	1.0426E-13	49285576.7	2.21274E-15	4.98498E-14 (45.06)	1.10637E-15 (1)
1.6403E+07	1.0708E-13	48632758.7	2.15450E-15	4.78950E-14 (44.46)	1.07725E-15 (1)
1.5982E+07	1.0990E-13	48005213	2.09926E-15	4.60647E-14 (43.89)	1.04963E-15 (1)
1.5583E+07	1.1272E-13	47401349.8	2.04678E-15	4.43482E-14 (43.33)	1.02339E-15 (1)
1.5203E+07	1.1554E-13	46819716.1	1.99686E-15	4.27356E-14 (42.80)	9.98428E-16 (1)
1.4841E+07	1.1835E-13	46258980.7	1.94931E-15	4.12185E-14 (42.29)	9.74656E-16 (1)
1.4496E+07	1.2117E-13	45717921.4	1.90398E-15	3.97810E-14 (41.80)	9.51990E-16 (1)
1.4166E+07	1.2399E-13	45195413.7	1.86071E-15	3.84403E-14 (41.32)	9.30354E-16 (1)
1.3851E+07	1.2681E-13	44690421.2	1.81936E-15	3.71661E-14 (40.86)	9.09679E-16 (1)
1.3550E+07	1.2963E-13	44201986.6	1.77981E-15	3.59608E-14 (40.41)	8.89903E-16 (1)

Table 3.6.6-2: Planck's Energy (E^P = h.f)

Note: the above (error) is a ratio and therefore a value of 1 represents zero error

The error when using Planck's constant 'h'

No error when using the modified version of Planck's constant 'h''

Max Planck gave us the crucial link to resolve Isaac Newton's *gravitational* constant (G):

These were Planck's constants; time, length and *mass*:

$t = \sqrt{[\hbar.G / c^5]} = 5.39096122598358E\text{-}44$ s

$\lambda = \sqrt{[\hbar.G / c^3]} = 1.61616952231127E\text{-}35$ m

$m = \sqrt{[\hbar.c / G]} = 2.1765500017459E\text{-}08$ kg

from which, it was possible to complete Planck's atom by defining the associated atomic energy and force formulas; E^P & F^P:

$E^P = \sqrt{[\hbar.c^5 / G]}$

$F^P = c^4/G$

But Planck did not realise that he had part-defined a fictitious proton-electron pair (chapter 2A.8.2) which led to the solution for 'G':

$G = \lambda^2.c^3/\hbar = \mathbf{6.67359232004333E\text{-}11}$ m^3 / kg.s^2

from which we can construct the formula for 'G' using Planck's units:

$G = a_o.c^2 / m_u = \mathbf{6.67359232004333E\text{-}11}$ m^3 / kg.s^2

m_u is the unit mass of a ultimate density (Table A2b)

There are three key electron orbital (shell) radii according to Planck (refer to Table 2.6.6-1), each of which relates to an associated electron velocity and temperature:

	Orbital Radius	**Orbital Velocity**	**Temperature**
Maximum R	$R_o = R_n.\zeta_v{}^2$	$v_o = c.\sqrt{[R_n/R_o]}$	$T_o = X.v_o{}^2$
Mean R	$R_m = R_n.\zeta_v$	$v_m = \sqrt{[v_o/c]}$	$T_m = X.v_m{}^2$
Minimum R	R_n	c	$T_n = X.c^2$
Table 3.6.6-3: The Planck Atom			

The resultant temperatures are:

Radon Gas Transition: $T_o = 210.19332853584$ K

This is the temperature below which the electrical charges in a proton-electron pair are identical ('e' = e'; refer to Chapter 3.6.7).

Mean: $T_m = 361962.55467156$ K

Neutronic: $T_n = 623316124.71718$ K

The highest possible temperature that can be generated naturally

3.6.7   e, e', e_n

The electrical charge in an electron is 1.60217648753E-19 C and designated the symbol 'e' (elementary charge unit). This charge is invariable in an electron, but not in a proton:

The static charge in a lone proton is always the same as that in an electron (e), but when a proton attracts an orbiting electron, its charge will increase (e'). This is facilitated by the proton's surplus non-polar magnetic capacity (*mass*). I.e.:

The electrical charge in a lone proton (and in an electron) is always; e

The electrical charge in a proton with an orbiting electron rises to; e'

The charge generated in a proton by its orbiting electron may be defined as follows:

$e' = m_p.RC . T/T_n$

which is also:

$e' = m_p.RC. (v/c)^2$

e' becomes e_n occurs immediately before a proton unites with its electron to become a neutron:

$e_n = m_p.RC = 2.94183820093364E-16$ C

It may be concluded, therefore, that whilst 'e' is used by the proton to maintain its partnership with its orbiting electron, 'e'' is used to repel adjacent atoms in matter.

Note: $e^2/4\pi$ is what is commonly referred to as the 'fine structure constant' (α)

'e' is one of the four primary constants (Table A1)

3.6.8 R_γ, R_∞, a_0

What is commonly regarded as *Bohr's* radius (a_0) is the orbital radius of what Bohr referred to as; an electron's *ground state*. But given that …

$a_0 = 4\pi.\varepsilon_0.\hbar^2 / m_e.e^2 = (h / 2\pi.m_e.c)^2 / R_n$
which breaks down to:
$a_0 = R_n.(\xi_v/4\pi)^2 = 5.2917721067E\text{-}11 \text{ m}$

… and that a 'ground state' can reasonably be said to occur at the point an electron ceases to provide sufficient kinetic energy to maintain its orbit, this will be when the electron is orbiting at its maximum orbital radius (minimum temperature).

But, given that Planck's minimum radius; $R_0 = 8.35643156381579E\text{-}09\text{m}$ occurs at a lower temperature; $\underline{T} = X_R/R_0 = 210.193328535837 \text{ K}$
it is reasonable to assume that 'a_0' does not represent a *ground state*.

Rydberg generated the following formula for his first constant:
$R_\infty = m_e.e^4 / 8.\varepsilon_0^2.h^3.c = 10973726.9561356 \ \{/m\}$
which breaks down to:
$R_\infty = \sqrt{R_n} / 4\pi.a_0^{1.5} \ \{/m\}$
which breaks down to:
$R_\infty = 1 / a_0.\xi_v = 10973726.9561356 \ /m$ (Table A2a)
and is his wave number (for electro-magnetic energy).

Rydberg generated the following formula for his universal energy constant for an electron:
$R_\gamma = R_\infty.h.c.(Z.n)^2 = 2.17987197684933E\text{-}18 \text{ J}$
which breaks down to:
$R_\gamma = R_n/a_0 \ . \ \frac{1}{2}.m_e.c^2 = 2.17987197684933E\text{-}18 \text{ J}$
and it occurs when the orbital radius of an electron is equal to; 'a_0'.

Therefore, it was Johannes Rydberg that defined 'a_0' (not Bohr), and this constant should be referred to as *Rydberg's radius*.

But Planck's maximum orbital radius is; $R_0 = a_0.(4\pi)^2$ (Table A2b)
And if Rydberg had modified his formula to reflect Planck's value:
$R_\gamma = \frac{1}{2} \ . \ (R_n / a_0.(4\pi)^2) \ . \ m_e.c^2 = 1.38042005551962E\text{-}20 \ \{J\}$ (Table A2a)
Rydberg would have revealed Planck's minimum electron kinetic energy 40 years earlier.

The relevance of Rydberg's radius (a_0) to EME is not yet understood, because there is no such thing as 'rest-mass' for electrons as they never come to rest at any temperature. I've no doubt it will come to light given time. However, this orbital radius occurs at a proton-electron pair temperature of 33192.4000063507 K, which represents the EME radiation at the margin between ultraviolet and X-Ray, i.e. the limit of visible light ($\lambda = 4.55633716784276E\text{-}08$ m & $f = 6.57968117714912E\text{+}15$ Hz). It must be noted that Rydberg's radius is fundamental to the creation of neutrons (chapter 3.6.15), and therefore plays a vital role in the workings of the universe.

3.6.9 R_s

Schwarzschild's radius is the radius of a spherical body, the non-polar magnetic (potential) energy of which, is sufficient at its surface to trap an electron travelling at light-speed (*photon*).

It is calculated as follows:

$R_s = 2.G.m/c^2$
where 'm' is the *mass* of the spherical body.

The following examples give some idea of the Schwarzschild radii of various bodies:

The minimum sized black-body of iron density (ρ_i):
$R_s = c.\sqrt{[\ 3\ /\ 8.\pi.G.\rho_i\ ]} = 1.42875013455622E+11$ m
$m = {}^4/_3\pi R_s{}^3$. $\rho_i = 9.6237854E+37$ kg

Body of matter of ultimate density (ρ_u):
Body radius: $r = c.\sqrt{[\ 3\ /\ 8.\pi.G.\rho_u\ ]} = 47494.1512680647$ m
Body mass: $m_s = {}^4/_3\pi.r^3.\rho_u = 3.19809876372352E+31$ kg
Proving that a proton, which has a much smaller mass than 'm' above, cannot trap a *photon* through *gravitational* force (magnetic charge) alone.
It is interesting to note that; $4\pi\ /\ (\varphi.m_s) = k$ (Coulomb's constant)

A proton:
$R_s = 2.G.m/c^2 = 2.48396784934951E-54$ m
Also proving that a proton cannot trap an electron through potential force alone. But if we apply the coupling ratio to Newton's force formula, Schwarzschild's radius is;
$R_s = 2.G.m\ /\ \varphi.c^2 = 5.63587590767792E-15$ m
i.e. exactly twice the neutronic radius ($R_n = 2.81793795383896E-15$ m)

However:
The Schwarzschild radius (R_s) is fictitious because an electron in free-flight cannot travel at velocity 'c' and there are no such things as *photons*.

3.6.10 R_n, t_n

The neutronic radius is the orbital radius of an electron when it is travelling at velocity; 'c'. At this speed, the electron and proton combine to become a neutron.

$R_n = \mu_o.e.RC$ { kg.m / C^2 . C . C/kg = m }

$R_n = G.m_p / \varphi.c^2$ { m^3 / kg.s^2 / kg . s^2/m^2 = m }

$R_n = \mu_o.e.RC = G.m_p / \varphi.c^2 = 2.81793795383896E-15$ m

The *gravitational* acceleration in the proton-electron pair according to Newton and Coulomb at this time:

$g_n = c^2/R = 3.18940728807838E+31$ m/s^2

$g_n = G.m_p / \varphi.R_n^2 = 3.18940728807838E+31$ m/s^2

The time an electron takes to orbit its proton when travelling at '*light-speed*' is;

$t_n = 2\pi.R_n/c = 5.90596121302193E-23$ s

'R_n' & 't_n' are two of the four primary constants (Table A1)

3.6.11   RAC, RAM

Relative atomic charge and relative atomic *mass* define the capacity of a particle to hold an electrical charge (e) or magnetic charge (*mass* 'm') respectively.
They are related as follows:

RAM {g/mol}

RAC {C/mol}

Divide either of the above by Avogadro's constant (N_A), you will get the capacity (m, e) of the particle {g, C}

Divide the ideal gas constant (R_i) by either of the above and you will get the *specific* capacity (c, q) of the particle {J/g/K, J/C/K}

Multiply the capacity by the *specific* capacity you will get the *relative* capacity (C, Q) of the particle {J/K} – Boltzmann's constant

3.6.12 N_A

Avogadro's number is the number of C^{12} (pure carbon) atoms in 12g and is recognised as; $N_A =$ 6.02214129E+23 /mole

However, one atom of pure carbon-12 has a *mass* of:
$m_C = 6.(m_e+m_p+m_n) = 2.00823909216E-23$ g

i.e.; $N_A = 1/(m_e+m_p)$ {/mol}
where m_e & m_p are specified in grams

and 12 grams of pure carbon-12 contains:
$N_A = 12/m_C = 5.97538412973187E+23$
which is 0.7764208% less than Avogadro's number

If corrected, this would, of course alter a number of constants such as:
R_i; X; X_R; c & q

However, throughout this book, all property values have been based upon Avogadro's value; *Avogadro's Number* has been left as *he* defined it.

The Mole:

A mole of matter (e.g. Quanta, atoms, molecules, etc.) is the mass that contains Avogadro's number (N_A) of particles of that matter. It also happens to be the RAM of that particle in grams.

E.g. a mole of water (H_2O) is 18.01528g (2x1.00794 + 15.9994)

3.6.13 k_B, R_i

Boltzmann's constant (k_B) defines the kelvin temperature scale based upon the potential energy of a proton-electron pair. He defined this scale thus:

k_B = 1.38065156E-23 J/K

which means that at the neutronic condition, the temperature of the proton-electron pair is:

$T_n = m_e.c^2$ / $Y.k_B$ = **6.23316124717170E+08 K**

If we wish to alter the temperature-scale we simply need to select our preferred neutronic temperature; say 1.0E+08 K, and recalculate his constant like this:

$k_B = m_e.c^2$ / $Y.T_n$ = **8.60582379963926E-23 J/K**

Together with Avogadro's constant, Boltzmann's constant defines the ideal gas constant:

$R_i = N_A.k_B$ = **8.24992342031355 J/K/mol**

3.6.14   K & h

This constant of proportionality is common to all orbits.

In elliptical orbits, its value is governed by the force-centre's *mass* and its formula is not only;

$$K = t^2/a^3 = (2.\pi)^2 / G.m_1$$

but also;

$$K = 2\pi/v_{max} \cdot 2\pi/v_{min} / a \ \{s^2/m^3\}$$

Where:
v_{max} = maximum orbital velocity (@ orbital perigee)
v_{min} = minimum orbital velocity (@ orbital apogee)
a = half the length of the orbital major axis

In circular orbits, its value is governed by the *satellite's mass*, and its formula alters to:

$$K = (2\pi/v)^2 / R \ \{s^2/m^3\}$$
Where:
v = orbital velocity
R = orbital radius

Its value can be any number greater than 0

For a proton-electron pair: $K = 0.15587874533403 \ \{s^2/m^3\}$

For our lunar system: $K = 9.91826542816423E-14 \ \{s^2/m^3\}$

For our solar system: $K = 2.97491436434708E-19 \ \{s^2/m^3\}$

For our Milky Way: $K = 3.35025744566253E-30 \ \{s^2/m^3\}$
Assuming an orbital eccentricity of 0.015941744 for our solar system

For our solar universe: $K = 5.66976153229753E-35 \ \{s^2/m^3\}$

Newton's constant of motion of an elliptical orbit is calculated like this;

$$h = R.v$$

and for universal expansion, it is calculated like this;

$$h = R.u/2$$

3.6.15 ξ_v, ξ_m

The **static ratio** defines the relationship between the *mass* of a proton and the *mass* of an electron, but because both have the same density, it also describes their relative volumes:

$$\xi_m = m_p/m_e = V_p/V_e = 1836.15115053207 \qquad \text{no units}$$

This value is very specific. It defines the orbital instant when magnetic field energy generated in the proton-electron pair equals the centrifugal energy in the electron. Which occurs when the electron is orbiting at the velocity of electro-magnetic energy (c) and simultaneously achieving the neutronic radius (R_n).

If ξ_m > the above value; the orbiting electron would impact its proton partner before it achieved 'c'

If ξ_m < the above value; the orbiting electron's velocity would achieve 'c' before it achieved R_n and thereafter remain constant

In both cases, neutrons could not be created and our universe would not exist.

The **dynamic ratio** defines the relationship between '*light-speed*' velocity (c) and Planck's minimum velocity (v_o):

$$\xi_v = c/v_o = 1722.0458764934 \qquad \text{no units}$$

It was found from:
Planck's constant; $h = \sqrt{(\pi.m_e.e^2.a_o / \varepsilon_o)} = 6.62607174469163E\text{-}34 \text{ kg.m}^2/s$
and;
Rydberg's constant; $a_o = \varepsilon_o.(h/e)^2 / \pi.m_e = 5.2917721067E\text{-}11$ m
But is derived from:
$$\xi_v = 4\pi.\sqrt{[a_o/R_n]} = 1722.0458764934$$

It has since been discovered that every property of the proton-electron pair is in some way related to the dynamic ratio:

	Minimum (m_1):	(m_2) Mean (m_1):	(m_2) Maximum:	units	formula:	M:M
T	210.19332853584	361962.554671561	623316124.717178	K	m_2/m_1	ξ_v
R	8.35643156381571E-09	4.85261843362263E-12	2.81793795383896E-15	m	m_1/m_2	ξ_v
v	174090.866621084	7224342.80705004	299792459	m/s	$\sqrt{[m_2/m_1]}$	ξ_v
G	3.62686268767106E+18	1.07552509449653E+25	3.18940728807838E+31	m/s^2	$\sqrt{[m_2/m_1]}$	ξ_v
T	3.01595419916531E-13	4.22043937529269E-18	5.90596121302193E-23	s	$\frac{2}{3}\sqrt{[m_1/m_2]}$	ξ_v
KE	1.38042005551962E-20	2.37714666443636E-17	4.09355561131267E-14	J	m_2/m_1	ξ_v
PE	-2.76084011103925E-20	-4.75429332887272E-17	-8.18711122262534E-14	J	m_2/m_1	ξ_v
F	-3.30385056103851E-12	-9.79737721789819E-06	-2.90535538991261E+01	N	$\sqrt{[m_2/m_1]}$	ξ_v
H	1.45477841280E-03	3.50569790763E-05	8.44796548491E-07	m^2/s	$[m_1/m_2]^2$	ξ_v
F	3.31570021944218E+12	2.36942154851033E+17	1.69320448260839E+22	Hz	$\frac{2}{3}\sqrt{[m_2/m_1]}$	ξ_v
λ	9.04160325599145E-05	1.26525589837942E-09	1.77056263481047E-14	m	$\frac{2}{3}\sqrt{[m_1/m_2]}$	ξ_v
V	0.17231810182	296.73967667583	510999.3366116	J/C	m_2/m_1	ξ_v
I	2712.81241061556	2712.81241061556	2712.81241061556	C/s	m=m	1
R	6.35200949172824E-05	1.09384517526775E-01	1.88365157359204E+02	J.s/C^2	m_2/m_1	ξ_v
Proton-Electron Pair Properties and the Static Ratio						
M:M refers to both 'mean:minimum' and 'maximum:mean' ratios						

3.6.16   B, RC

The magnetic field 'B' as described by Lorentz is actually $1/RC$, where RC is the relative charge capacity of Quanta (Table A3). And the orbital radius at which an electron and a proton combine to create a neutron may be calculated using it:

$R_n = \mu.e/B$ (chapter 3.6.10)

Lorentz's magnetic field constant was originally derived as follows:

$B = \mu_o.I / 2.\pi.R$
where: $R = 2.R_n$ & $I = e$

But can also be calculated thus:

$B = 1/RC = m_e/e = 5.685634367312E\text{-}12$ {kg/C}

RC is the relative [electrical] charge capacity of matter of ultimate density (i.e. Quanta), but may also be applied to any *mass*; you simply need to rationalise its density:

Relative charge capacity of planet earth; $RC_E = e/m_e . \rho_E/\rho_u$

3.6.17   X, X$_R$

X & X$_R$ are both [heat] energy transfer coefficients for the electron.

$T = PE / k_B.^3\sqrt{[\frac{1}{2} . \sqrt{[(4\pi)^2.a_o / R_n]}]}$
Which leads to; $T = X.v^2 / e^2$
and because 'e' is a constant; $T = X.v^2$

$X = T_n/c^2 = 6.93532716478941E\text{-}09 \ K.s^2/m^2$

$X_R = R_n.T_n = 1.75646616508035E\text{-}06 \ K.m$

X {K.s^2/m^2} is an energy transfer coefficient for an electron. As an orbiting electron receives electro-magnetic energy, it converts this energy into velocity according to the relationship:

$v = \sqrt{[T/X]}$

X$_R$ {K.m} is also an energy transfer coefficient for an electron. This coefficient is another way of writing X but converting electro-magnetic energy into orbital radius:

$R = X_R/T$

Moreover, if:
$T = X.v^2 = X_R/R$
$R = X_R / X.v^2$
At light-speed (c)
$R_n = X_R / X.c^2 = 2.81793795383896E\text{-}15 \ m$

According to the relationship: $T = X.v^2$
$T_n = X.c^2 = 623316124.71718 \ K$
which is achieved by a proton-electron pair at the instant of their union as a neutron, it therefore represents the highest possible temperature achievable by natural means.

This is the actual temperature at the centre of all stars.

You may have noticed that; $T = X.v^2 / e^2$ is similar to Newton's *gravitational* force; $F = G.m_1.m_2 / R^2$, Coulomb's force; $F = k.q_1.q_2 / R^2$, and Gilbert's and Maxwell's formulas for force and energy. It is therefore anticipated that all of these formulas will eventually become just two; one for magnetic charge (*mass*) and the other for electrical charge.

An interesting relationship for the above heat constants is as follows:
$(2\pi)^2.X/X_R = K$ {s^2/m^3}

Where: **K** is Isaac Newton's orbital constant of proportionality for circular orbits (for the atom):
$K = t^2/a^3 = 0.15587874533403$ {s^2/m^3}

3.7 The Magical Constants

Watch this, its magic!

1) Let me ask you a few questions:

Do you believe that Isaac Newton and Charles-Augustin de Coulomb understood their own force-laws?

Do you accept the Magnetic Constant?
$\mu_o = 1E\text{-}07$ kg.m / C^2

Do you accept the Permittivity Constant?
$\varepsilon = e^2 / 4\pi.h' = 8.85418775855161E\text{-}12$ C^2 / J.m

2) If your answer to all of the above is yes, then you must also accept the coupling ratio:

Magnetic: $F_m = G.m_1.m_2 / R^2$ (Newton & Gilbert)

Electrical: $F_e = k.q_1.q_2 / R^2$ (Coulomb & Maxwell)

$\varphi = F_m / F_e = G.m_p.m_e / k.e^2 = 4.40742111792334E\text{-}40$

3) Newton's force-formula tells us that; $g = v^2/R = G.m/R^2$

From which, we get, when replacing 'v' with 'c':

$R = G.m_p / \varphi.c^2 = 2.81793795383896E\text{-}15$ m (R_n!)

So, we know that anything orbiting a proton at 'c' must be orbiting at 'R_n'

'R_n' is therefore a constant that represents the orbital radius of an electron travelling at 'c'.

Newton and Coulomb both said so!

4) The Magnetic Constant:

$\mu_o = 1E\text{-}07$ kg.m/C^2

but what is 1E-07?

$m_e.R_n/e^2 = 1.000000000000000E\text{-}07$ kg.m/C^2 Coincidence?

5) Let's just check using Coulomb's constant:

$k = \mu_o.c^2 = 8.98755184732667E+09$ kg.m^3 / C^2.s^2

Incorporating 4) above (μ_o) we find:

$k = m_e.c^2.R_n / e^2 = 8.98755184732667E+09$ kg.m^3 / C^2.s^2 !

6) So, now we can find the significance of Coulomb's constant:

$k = (m_e.c^2) . R_n / (e^2)$ {kg.m^3 / C^2.s^2 = kg.m^2/s^2 . m / C^2 = J.m/C^2}

Remember 'J.m/C^2' when it comes to Planck's constant (h)

7) Now let's check using Planck's constant:

$h = \sqrt{[\pi.m_e.a_o.e^2 / \varepsilon]} = 6.62607174469163E-34$ kg.m^2/s

Planck claimed its units as 'J.s'

However, a frequency ratio must be applied to his formula to achieve these units:

$h = \sqrt{[\pi.m_e.a_o.e^2 / \varepsilon]} . [f_1/f_2]$ {kg.m^2/s . s/s} (J.s)

8) If we expand ε_o in Planck's formula:

$h = \sqrt{[4.\pi^2.m_e^2.c^2.a_o.R_n]} = 6.62607174469163E-34$ kg.m^2/s

which can be reorganised thus:

$h = \sqrt{[a_o.(4.\pi)^2 . R_n]} . \frac{1}{2}.m_e.c$ #

$= 6.62607174469163E-34$ kg.m^2/s

i.e. a mean radius x kinetic energy! [but there is something missing #].

9) So let's change Planck's formula a bit:

$h' = \sqrt{[a_o.(4.\pi)^2 . R_n]} . \frac{1}{2}.m_e.c . v$ {kg.m²/s . m/s}

but what is v?

Let's try; $v_o = c . \sqrt{[a_o.(4.\pi)^2 . R_n]} = 174090.866621084$ m/s

$h' = 1.15353857232684E-28$ kg.m³/s² {J.m}

but now the units work.

10) h' also equals ...

$h' = \frac{1}{2}.m_e.c^2.R_n$ {J.m}

validating the modified version of Planck's constant (remember Coulomb's constant J.m/C²)

In this case; 'R_n' is not only the orbital radius of the electron when travelling at 'c', it is also the coincident electro-magnetic amplitude

(which is the purpose of Planck's constant).

11) h vs h' in the atom:

Now if we calculate both constants to find the potential energy of an orbiting electron; KE^P using h & PE using h' ...

Whilst it appears that Planck's constant (h) doesn't work; h' does!

T	R (A)	v	KE	E^P (error)		E (error)	
6.2332E+08	2.8179E-15	299792459	4.0936E-14	1.12193E-11 (274.1)		4.09356E-14	
3.1166E+08	5.6359E-15	211985281	2.0468E-14	3.96662E-12 (193.8)		2.04678E-14	
2.0777E+08	8.4538E-15	173085257	1.3645E-14	2.15915E-12 (158.2)		1.36452E-14	
1.5583E+08	1.1272E-14	149896230	1.0234E-14	1.40241E-12 (137.0)		1.02339E-14	
1.2466E+08	1.4090E-14	134071264	8.1871E-15	1.00348E-12 (122.6)		8.18710E-15	
1.0389E+08	1.6908E-14	122389759	6.8226E-15	7.63376E-13 (111.9)		6.82260E-15	
8.9045E+07	1.9726E-14	113310899	5.8479E-15	6.05785E-13 (103.6)		5.84795E-15	
7.7915E+07	2.2544E-14	105992640	5.1170E-15	4.95828E-13(96.90)		5.11695E-15	
6.9257E+07	2.5361E-14	99930819.7	4.5484E-15	4.15529E-13 (91.36)		4.54840E-15	
6.2332E+07	2.8179E-14	94802699.6	4.0936E-15	3.54785E-13 (86.67)		4.09356E-15	
5.6665E+07	3.0997E-14	90390827.4	3.7214E-15	3.07522E-13 (82.64)		3.72141E-15	
5.1943E+07	3.3815E-14	86542628.5	3.4113E-15	2.69894E-13 (79.12)		3.41130E-15	
Planck's Energy (E^P = h.f)							
Note: the above (error) is a ratio and therefore a value of 1 represents zero error							

But we have established a definite link between Coulomb's and Planck's constants.

12) Comparing Planck's minimum velocity (v_0) with light-speed:

$\xi_v = v_0 : c$

which we can use (ξ_v) to simplify Rydberg's wave number ...

$R_\infty = m_e.e^4 / 8.\varepsilon^2.h^3.c = 10973726.9561359$ /m

which becomes:

$R_\infty = 1 / a_0.\xi_v = 10973726.9561359$ /m

13) Now if we apply Rydberg's wave number; R_∞ ...

... to his own energy constant:

$R_\gamma = R_\infty.h.c.(Z.n)^2 = R_n/a_0 . \frac{1}{2}.m_e.c^2 = 2.17987197684936E\text{-}18$ J

By an amazing coincidence, it just happens to be the kinetic energy of an electron orbiting at 'a_0' metres $\{T = 33192.4K\}$.

So, Rydberg was telling us that he defined an electron's rest-state to occur when orbiting at 'a_0'.

14) Rydberg (not Bohr) therefore defined 'a_0':

$a_0 = R_n . \frac{1}{2}.m_e.c^2 / R_\gamma \{m\}$

So why do we refer to this dimension as Bohr's radius?

Because Bohr originally declared it to be the orbiting radius of a hydrogen electron at rest-mass; the only electron property he managed to calculate. He subsequently declared that electrons do not orbit (Quantum Theory) ...

... whereas Rydberg had actually told us that the orbital radius of an electron can be found from:

$R_e / R_n = \frac{1}{2}.m_e.c^2 / KE_e$

which has subsequently proved to be correct, therefore Bohr was wrong, so we should be referring to 'a_0' as the Rydberg radius.

15) Planck's mean [orbital] radius can be found thus:

$R_m = R_n.\zeta_v = R_n / a_o.R_\infty$

Having already established that ...

$h' = \frac{1}{2}.m_e.c^2 . R_n$ {J.m}

and:

$R_\infty = 1 / a_o.\zeta_v$

... we have now established a definite working link between Planck's and Ryberg's constants.

16) Coulomb's constant is related to h' thus:

$h' = \frac{1}{2}.e^2/\varepsilon_o = k.e^2 = 1.15353857232684E-28$ {J.m}

It is interesting to note that using the modified version of Planck's constant (h') we can find the fine-structure constant 'α':

$2.h'.\varepsilon = e^2/4\pi = 2.04272942122269E-39$ C^2

17) Electrical and magnetic forces must be equal when the electron is orbiting at 'c'

i.e. magnetic force = electrical force @ R_n

$F = R_n^2.m_e.c^2/R_n^3 = k.e^2/R_n^2$

$R_n = k.e^2 / m_e.c^2 = 2.81793795383896E-15$

$\{kg.m^3.C^2 / C^2.s^2 / (kg.m^2/s^2) = m\}$

Confirming again that Coulomb said so.

18) and finally ...

$k_B = m_e.c^2 / Y.T_n$

i.e. the potential energy in a proton-electron pair at the time of its conversion to a neutron ...

... confirming that electrons orbit protons in circular paths.

Summary

The above sequence demonstrates an unequivocal connection between Rydberg's, Planck's, Newton's, Coulomb's, etc. constants and formulas through energy, and confirms that the neutronic ratio (R_n) is a fundamental constituent of all of them.

To Conclude

'R_n' was initially discovered long-hand (iteratively). Since then, I have not only established an unequivocal link between Newton's, Coulomb's, Planck's and Rydberg's constants, but also that they are all dependent upon this neutronic radius.

Therefore, 'R_n' must be a real value that not only complies with known and accepted physical constants, it also validates those formula's in terms of their units and explains their meaning.

If we accept the validity of 'R_n' we must also accept the concept of circular orbits in atoms.

If we accept circular orbits in atoms, we must reject Quantum theory.

Moreover, because; μ_o, ε_o, k, h, h', R_∞, R_γ are all dependent upon 'R_n'

Relativity must be wrong because it is based upon the photon, and 'R_n' confirms that electrons cannot leave an atomic shell when orbiting at 'c'; i.e. electrons cannot travel in free flight at velocity 'c'. Therefore, photons cannot exist!

And, according to Relativity:

$v = v / \sqrt{[1+v/c]}$!

which means an electron could never achieve 'c'

($c = c/\sqrt{2}$); so 'R_n' would be impossible

{$c = c/\sqrt{2}$ is of course nonsense because $c \neq c/\sqrt{2}$}

The above sequence of calculations pulls all the primary constants together and finally proves that Newton and his fellow pre-twentieth century physicists were correct.

The PHYSICAL CONSTANTS

4 The Laws of Thermodynamics

The First Law of Thermodynamics: Conservation of energy
Energy can never be lost, it can only be transformed or transferred.

The Second Law of Thermodynamics: Heat will not spontaneously pass from a colder body to hotter body

A high-energy source (hotter body) will spontaneously lose energy to a low-energy source (colder body) but you must add work if you want energy to transfer in the other direction (up-hill so to speak). This law essentially states that it is impossible to create energy from nothing.

This law also claims that energy can, and in fact is, lost by a system to its surroundings but that the reverse cannot happen i.e. an increase in disorder is an inevitable feature of time.

The Third Law of Thermodynamics: The entropy of a substance approaches zero as its temperature approaches zero (absolute)

Entropy is the term used to define disorder. The higher a substance's temperature the more disordered will be its atomic structure and the higher its entropy. E.g. gas has a higher entropy than a solid substance.

The PHYSICAL CONSTANTS

Appendices

References, symbols, glossary, etc. used throughout this book along with a summary list of corollaries and hypotheses.

A1 General

N/A

A2 References

Most of the references used for the creation of this book are from original work supplied in CalQlata (www.calqlata.com), but some additional sources are listed below:

Magnificent Principia; Colin Pask; 978-1-61614-745-7

Seven Brief Lessons on Physics; Carlo Rovelli; 978-0-141-98172-7

Science Data Book; Open University; 0 05 002487 6

Science and Technology Dictionary; Chambers; 0-550-18026-5

A Dictionary of Scientific Units; H G Jerrard & D B McNeill; 0-412-28100-7

It is important to note here that most of the sources here are from work done by pre-20th Century scientists that are universally known and available from sources too numerous to mention here.

The principle sources for physics today, and those that apply to this book are listed below:

Philosophiæ Naturalis Principia Mathematica Rev. IV; Keith Dixon-Roche; 978-1-07215-605-5

The Atom; Keith Dixon-Roche; 978-1-08610-029-7

The Neutron; Keith Dixon-Roche; 978-1-08251-683-2

The Life & Times of the Neutron; Keith Dixon-Roche; 978-1-08239-479-9

The Universe; Keith Dixon-Roche; 978-1-70753-878-2

The Mathematical Laws of Natural Science; Keith Dixon-Roche; 979-8-61029-449-0

A3 Glossary

Atomic Particle	One of the three components that comprise an atom
Coupling Ratio (φ)	The ratio of the coupling force due to a magnetic charge and the coupling force due to an electric charge: $\varphi = G.m_p.m_e \div k.e^2$ (chapter 3.6.4)
EME	Electro-magnetic energy
Ultimate Density	The mass-density of all three atomic particles $\rho = 7.12660796350449E{+}16 \text{ kg/m}^3$ Nothing in nature has a 'mass'-density greater than this value
Viscous	Solid or liquid matter in which magnetic field energy is greater than an electron's electrical charge

All other definitions can be found on the following web page:
http://calqlata.com/help_definitions.html

A4 Symbols

Refer to Chapter 2 for a list of all the symbols used in this book.

The most prominent subscripts are listed below:

mass	e	electron
	p	proton
temperature	c	cold
	o	minimum Planck
	m	mean Planck
	n	maximum Planck
Rydberg	γ	energy constant
	∞	wave number
	o	orbital radius (a_o) *occasionally referred to as the Bohr radius*
Others	u	Ultimate
radii	n	Neutron orbit radius
	s	Schwarzschild radius
	1	force-centre
	2	satellite
energy	e	electron
	p	proton

The most prominent superscripts are listed below:

Force	N	Newton
	P	Planck

The PHYSICAL CONSTANTS

A5 Useful Formulas

Equidistant arc-length between 'n' points on the surface of a sphere:

$d = \pi.A / C.n$

where C is the circumference of the sphere

Linear distance across arc-length 'd' (above):

$\ell = 2.R.Sin(\frac{1}{2}.d/R)$

but if you know 'ℓ' and need to find 'n':

$n = \pi / Asin(\frac{1}{2}.\ell/R)$

and if ℓ=R:

$n = \pi / Asin(\frac{1}{2}) = \mathbf{6}$

Lorentz's Equation (magnetic force or field strength):

$F = q.v.B$

Which becomes:

$F = q.g.R.B$

for the laws of orbital motion

Where:

q is the total electrical charge = $q_1.q_2 / m_e.(q_1+q_2)$

v = relative velocity (electrical circuits)

g = gravitational attraction between m_1 & m_2

R = radial separation between m_1 & m_2

$B = \mu_o.e/R_n = R_n.m_e/e^2$. $e/R_n = m_e/e = 1/RC$ kg/C

RC is the relative atomic charge capacity of an electron {C/kg}

$B = 1/RC = 5.685634367312E-12$ kg/C

Inter-atomic force factor (F_T):

$T_k = T_n / \xi_m.Y^2$

$F_T = T_1/T_k$

T_1 = measured temperature of atom (shell-1 temperature)

The PHYSICAL CONSTANTS

A6 The Heroes

The heroes of this story, to which I offer my gratitude, are listed below

It is not necessary to identify the invaluable contributions made by each of these contributors, they are all widely known and available in almost every scientific publication in circulation today.

Nicolaus Copernicus (Polish) 1473-1543
William Gilbert (English) 1544-1603
Tyco Brahe (Danish) 1546-1601
Galileo Galilei (Italian) 1564-1642
Johannes Kepler (German) 1571-1630
Christiaan Huygens (Dutch) 1629-1695
Isaac Newton (English) 1642-1727
Edmund Halley (English) 1656-1741
Charles-Augustin de Coulomb (French) 1736-1806
Hans Christian Ørsted (Danish) 1777-1851
Michael Faraday (English) 1791-1867
Josef Stefan (Austria) 1815-1863
James Clerk Maxwell (Scottish) 1831-1879
William Crookes (English) 1832-1919
Ludwig Boltzmann (Austria) 1844-1906
Hendrik Lorentz (Dutch) 1853-1928
Jules Henri Poincaré (French) 1854-1912
Johannes Robert Rydberg (Swedish) 1854-1919
Max Karl Ernst Ludwig Planck (German) 1858-1947

The others that were instrumental in the completion of this book are:

My long-suffering wife (Brigitte) sub-editor and critic

My daughter (Eléonore), who initiated this project

Kenneth Pickering friend & editor, who first suggested that I write it

My thanks go out to all the above each of whom have provided a valuable piece of the puzzle without which the final solution would not have been possible, along with my sincere apologies to anybody I have unintentionally omitted.